V&R Academic

NATIONAL TAIWAN UNIVERSITY PRESS

Reflections on (In)Humanity

Volume 9

Edited by

Sorin Antohi, Chun-Chieh Huang and Jörn Rüsen

Hardy F. Schloer / Mihai I. Spariosu

The Quantum Relations Principle

Managing our Future in the Age of Intelligent
Machines

With 31 figures

V&R unipress

National Taiwan University Press

Published in cooperation with the Institute for Advanced Studies in Humanities
and Social Sciences, National Taiwan University.

This book series is sponsored by the Orbis Tertius Association, Bucharest.

Bibliographic information published by the Deutsche Nationalbibliothek

The Deutsche Nationalbibliothek lists this publication in the Deutsche Nationalbibliografie;
detailed bibliographic data are available online: http://dnb.d-nb.de.

ISSN 2198-5278
ISBN 978-3-8471-0662-3 (Print, without Asia Pacific)
ISBN 978-986-350-192-3 (Print, Asia Pacific only)

You can find alternative editions of this book and additional material on our website: www.v-r.de

This work was supported by a grant from the Romanian National Authority for Scientific Research,
CNCS – UEFISCDI, project number PN-II-ID-PCE-2011-3-0771.

"Imagination is more important than knowledge. For knowledge is limited to all we now know and understand, while imagination embraces the entire world, and all there ever will be to know and understand." — Albert Einstein

"It will, of course, be said that such a scheme as is set forth here is quite unpractical and goes against human nature. This is perfectly true. It is unpractical and goes against human nature. This is why it is worth carrying out, and that is why one proposes it. For what is a practical scheme? A practical scheme is either a scheme that is already in existence, or a scheme that could be carried out under existing conditions. But it is exactly the existing conditions that one objects to; and any scheme that could accept these conditions is wrong and foolish. The conditions will be done away with, and human nature will change. The only thing that one really knows about human nature is that it changes." — Oscar Wilde

Contents

Preface

The world is in turmoil all around us. For the past two decades, nation-states have been on the move, realigning their political orientations, associations, and systems of values and beliefs. At an increasingly fast pace, old countries and empires are collapsing, and new societies and nations are born. Unfortunately, war, ethnic and religious cleansing, torture, social and political injustice, unfair or oppressive commercial competition and irresponsible financial behavior on a global scale have also become part of this process of radical change.

Perhaps at no other time in human history have local and global leaders been more challenged than now to find the right path to a sustainable future for humankind. The lack of clean and affordable energy, the diminishing supply of clean water, the imminent global collapse of the food supply to a vastly expanding human population, the fast growth of pollution due to the increasing industrialization of large nations, climate change issues, ever-sharpening conflicts due to the unequal access of many communities to natural resources and affordable technological solutions, and the impasse of philosophical and ideological differences, all of these are factors that decision-makers are called on to accept as reality and, therefore, find solutions to.

To compound the problem, the effectiveness and speed of contemporary communication technologies and public media allow local and global leaders little time to catch their breath and carefully consider their actions. Momentous decisions must be made almost instantaneously, in real-time, often with unforeseen, devastating consequences for humanity. This escalation in speed and volume of flawed political, economic, and environmental decision-making has produced an avalanche of global instability and uncontrolled change, which has led, in turn, to more turmoil and confusion at all levels.

And yet, this global crisis can also be seen as a unique opportunity for further human development. Because several important shifts are now occurring simultaneously, they are creating a perfect time storm, so that current and future human actions can be radically transformed. In the last three decades alone, we have made tremendous progress in devising new, environmentally responsible technologies and novel methods of applying traditional ones. Biology, physics, chemistry and medicine lead this progress, side by side with information and communication technologies. We have all the tools at our fingertips to transform this planet into a sustainable global community.

We need to act now, however, and must be willing to develop and use these new tools not only in an intelligent, but also in a wise manner. We can no longer get by with partial patchworks of fixes and need to begin thinking in terms of interrelated and integrated scientific and technological strategies to repair the negative, unintended consequences of the modern age of industrialization and globalization. Above all, we need to embrace science and technology as tools for our higher, ethical development, instead of abusing them for destructive, competitive, and selfish or greedy purposes. In other words, we need also to be willing to undergo a spiritual transformation, a radical change in our present mentality.

We need not only to become educated about the consequences of our decisions, but also begin to understand the complicated interrelations and hidden dependencies of global reality. This complex, fast-moving reality requires a fully committed approach and willingness to investigate and understand the issues at hand, across all ideologies, philosophies, societies, and nations. Thus, the world in all its facets must become intercultural and transdisciplinary. This is as true for economical and political realities, as it is true for science, technology and education. Studies of problems, decisions, and implementations of policy need to be carried out through an integrated and transformative global partnership, based on full global awareness.

In short, we need to refocus and look at our planet as a multilevel, integrated system. Our global community needs to start functioning as a harmonious whole, much like a very complex super-organism, perpetually in a learning mode and flexible enough to make the vital adjustments and changes necessary for its continued wellbeing. Two previous books, *Global Intelligence and Human Development* (Spariosu 2005) and *Remapping Knowledge* (Spariosu 2006), have laid out the foundational principles and collective actions needed to set us on the road to building an intelligent planet. The present book, which can best be read in conjunction with the other two, continues to describe the kind of conceptual and technological tools needed to further human development toward global intelligence. We define global intelligence as the ability to understand, respond to, and work toward what is in the best interest of and will benefit all human beings and all other life on our planet. This kind of responsive understanding and action can only emerge from continuing intercultural research, dialogue, negotiation, and mutual cooperation; in other words, it is interactive, and no single national or supranational instance or authority can predetermine its outcome. Thus, global intelligence, or intercultural responsive understanding and action, is what contemporary nonlinear science calls an emergent phenomenon, involving lifelong learning processes.

The Quantum Relations Principle is precisely such an integrated conceptual and technological tool, designed to move us, humans, one step farther not only toward global intelligence, but also toward planetary wisdom. In Part One of the book, we lay out the scientific and philosophical foundations of the Quantum Relations Principle (QR for short). QR creatively combines the insights of the theory of relativity, quantum mechanics, general systems theory, and contemporary cognitive and social science, in order to bring together, in transdisciplinary fashion, the study of physical phenomena with that of mental and social phenomena. In computer science, QR is a revolutionary, theoretical and practical measuring tool that allows us to take into account, simultaneously, the experiences of multiple observers (humans and/or machines) of the same event or series of events. It helps us build dynamic models of reality that can measure the interactions of mutual causality among various observers and can make reasonably accurate assumptions about how they may behave and affect each other in the future.

In the first two chapters of Part One, we outline the scientific foundations of QR in relation to quantum physics and the theory of relativity. From quantum mechanics we borrow the idea of quanta (the smallest units or "elementary particles" of observed reality that cannot be further reduced within the theory), as well as the principle of uncertainty, extending them to the human mind and social phenomena. From Einstein's theory of relativity, we borrow the idea that objects carry their own frame of reference with them, and that their interactions (making a measurement, for example) are determined by the rules that govern those frames of reference. Thus, QR theory provides a description of the elements, structures, and interactions between objects as they evolve over time. The crucial difference from quantum physics and relativity theory, however, is that in Quantum Relations objects need not be physical. They can also be concepts, relationships, or sets of mental, emotional, and social objects. QR attempts to derive, for each such set, the rules that govern the interactions of the objects in the set, as well as the interactions of different sets.

In Chapter 3 we describe the basic principles of computation and the main computational features of Quantum Relations. QR revolves around two fundamental concepts that can equally be translated into the mathematical language of quantum mechanics and constitute the cornerstones of any technology platform based on the QR principle. These concepts are "data fusion objects" (DFOs) and "frames of reference" (FORs). Data fusion objects or DFOs can be defined as mental quanta or elementary particles. They interact according to well-defined rules, and the result of their interaction can equally become a computed function. DFOs arise within multiple FORs. Each FOR can be represented as a metric space, i. e., as a set of DFO elements, with one or more functions. Furthermore, a FOR can also be a DFO and vice versa, depending on their respective positions in the hierarchic space structure. Thus, a DFO can be an elementary particle in a higher-level FOR. In turn, this FOR can be a DFO of another, higher-level FOR structure, and so forth.

The DFO/FOR model has many technical advantages, including self-adapt-ability: it will automatically search for the best method and the shortest path to accomplish its goal. Even more importantly, the DFO/FOR model is capable of self-organization, because data and functions are implemented as sets of hier-archical objects. A FOR containing many DFO structures can also contain rules for the creation of new DFOs, the interaction between its DFOs, and the calcu-lation of functions between smaller DFOs, including the creation of new objects that embody certain relationships between these smaller DFOs. Thus, DFOs and FORs provide a natural model for general parallel computation. Since DFOs and FORs are discrete objects, they can be implemented on multiple processor systems, and calculations can be performed in parallel.

Furthermore, the DFO/FOR model is both modular and extensible. This means that a set of computations on one data set can be transformed into another data set and used by the second data set to define a third set of new functions, translating the preceding FOR into the new one. In addition, a FOR can contain rules for logical inference and deduction that operate on its com-ponent DFO objects. The fact that FORs are also considered DFOs for higher-level frames allows lower-level frames to define data properties. DFOs could equally be used to pose queries on other DFO frames. It can thus translate and incorporate any software program or computer language into its database, thus solving the currently intractable problem of systemic compatibility and inter-changeability in computer programming.

In addition, the DFO/FOR model is compact and adaptable, expressly de-signed to handle extremely large quantities of data, on the scale of gigabyte and terabyte sets, and to provide methods for manipulating them through parallel processing systems. The model can handle data storage, recuperation, and processing with great flexibility and practically no data loss. It assumes that no piece of information or knowledge from its database can ever become obsolete, because it may always turn out to be relevant in a different DFO/FOR config-uration or correlation between data sets.

Finally, in the last chapter of Part One, we place the Quantum Relations Principle in a larger philosophical context and point out its global ethical im-plications. QR shares the theoretical advantages of general systems theory and its offshoots, the theories of complexity and self-organization, over their sci-entific, reductionist counterparts, especially within a global reference frame. The DFO/FOR model is based on the "web of life" (Capra 1997) in its most diverse and complex aspects, including human relations and interactions. Unlike most reductionist scientific theories, QR implicitly acknowledges diversity and alterity as the very conditions of existence. It can take into account and process widely different cognitive perspectives, including linguistic, philosophical, cultural, sexual, and other observer-dependent variables.

Like other contemporary strands of systems theory, QR acknowledges that hierarchies as modes of organization are best understood not as "centers of command and control," but as reference frames or levels of complexity embedded or nestled within each other and engaged in constant communication and mutual interaction. QR thus supports and enhances a cooperative, symbiotic view of our universe, in which all living and nonliving components of the global system and subsystems depend on each other for their well-being and in which each perspective needs to be acknowledged and respected as potentially valuable for the common good.

Another theoretical advantage of QR within a global reference frame is that it shares the nonlinear views of the ancient tradition of wisdom, or the *philosophia perennis* as Leibnitz called it. We briefly examine such concepts as mutual causality or dependent origination, amplifying feedback loops, resonance, and the web of life as they first appeared in early Buddhism and Daoism and are now shared by the contemporary philosophy of process and general systems theory. Thus, these concepts constitute an excellent philosophical and scientific meeting point not only of West and East, but also of North and South, because they can be found, in one form or another, virtually in all known civilizations on our planet. They have gained great relevance in our age, because the planetary framework of globalization implies different rules and principles of human interaction, which need equally to be remapped and reorganized at the local and regional levels.

Furthermore, human evolution itself has revealed the wisdom of the old sages who advised the rulers of their time to move from a warlike mentality and violent competition to mindful, peaceful and innovative coexistence with one another and with all other beings on our planet. The QR principle is in full consonance with this mentality of peace, which also implies a second-order ethical system. Unlike first-order ethics, in which one accords preferential treatment to one's own group, be it family, tribe, nation, religion, etc., the second-order ethics is based on the Golden Rule (treat all human and other beings as you would like to be treated yourself) and is, therefore, the most appropriate mode of thought and behavior to adopt within a planetary reference frame.

In Part Two of the book, we propose concrete, QR technology-supported, global solutions, in line with the QR principle and the ethical precepts of the *philosophia perennis*, in four essential fields: big-data mining and processing, commerce, healthcare, and global learning and research. In Chapter 5, we describe one of our most exciting QR applications, an interactive content search and delivery engine, which we have called Q-Search and which renders obsolete the current Google and Yahoo/Microsoft search and transaction technologies. Indeed, it can revolutionize the whole field of intelligent, interactive management of information and content in real-time.

Q-Search is a problem-solving technological platform that operates broadly in three domains relevant to search issues. It excels at: 1) real-time content acquisition; 2) sophisticated real-time data processing with an emphasis on linguistic content analysis; and 3) on-demand query, computation, and delivery. In addition, Q-Search offers advanced monitoring and management tools to keep data centers aware of and provide flexibility in information, product and content offerings.

The most significant innovation of Q-search, however, is that it moves far beyond text and numbers. Unlike current search engines, which are text-based and still mostly static, Q-search can, for example, store and retrieve a moving image, a sound file, or a sequential recording of behavior, as a computable object through its QR time-series representation (that is, representing data as a sequence of objects over some time base). The ability of storing and retrieving such data as computable objects, rather than as fixed images or character strings, completely transforms the browsing experience.

The "browser" of the future will be a virtual world in which the user searches, finds, manipulates, and transforms objects, and not a simple text-based, data retrieval system. Q-Search technology can easily build the platform for this virtual world of content search, delivery and consumption models. Even more importantly, Q-Search can remember, preserve and dynamically track and record user-based frames of reference in order to reproduce the "world of the user" inside computers and databases, including the user's unique behaviors and interactions with his or her environment. By continuously estimating the users' future behavior and comparing it against their actual interactions, the QR-based search system becomes highly predictive in relation to the individual user, as well as to the totality of all recorded or known users. This feature will give us very early knowledge of unfolding trends, conditions, evolutions and potential risks, both individually and systemically. Most importantly, a user may be a person, group of persons (even temporarily constituted), machine, network of machines, and any combination thereof. In other words, Q-Search incorporates the full effect and advantage that QR provides from an ontological, epistemological, ethical and practical vantage point.

We employ Q-Search technology in the next three QR global applications that we describe in the remaining chapters of our book. Thus in Chapter 6, we propose a fully automated, global, commercial transaction-space. We have called this space the Global Value Exchange (GVE), because all of its operations and transactions are based on actual products and services, exchanged at their real value on all the levels of commercial participation. The GVE system will be the first enabling technology that potentially unites all producers and consumers of any product and service, anywhere in the world, in one single, standardized, transaction space.

The main goals of the GVE are: 1) to reintroduce, in the emergent global economy, the fair exchange of actual goods and services of real value; 2) to create stability through the precise calibration of economic units and to remove different currencies or instruments that have different values in different places; 3) to remove the highly speculative edge and unnecessary intermediaries that charge exorbitant interest, leverage between creditors and debtors, and are generally oriented toward usurious practices. In this sense, the GVE automatically adopts and applies some of the basic principles of traditional commerce, based on fair and mutually advantageous trading and honestly earned profit.

In addition, GVE reduces economic inefficiency to a minimum and, therefore, produces the fairest value, taking into account all of the elements that unite a product or service offering with its consumer. Most importantly, for the first time in human history, the fundamental ethical principles belonging to the second-order ethics, which are the most appropriate within a planetary reference frame, will be inbuilt as automatic features of the global economic system. This system will be based on a truly free market, without governmental or any other outside intervention, and will be impervious to fraud, corruption, and abusive manipulation. It is self-organizing, self-governing and self-policing for the equilibrated benefit of all.

Thus, the GVE system can be seen as a result of a long evolutionary process and constitutes a new stage in human development, viewed within a global reference frame. Within the GVE system, the creation of products and services will become a co-evolutionary process, governed by natural feedback loops between producers and consumers. The GVE system will therefore generate a symbiosis between those who produce actual goods and services and those who ultimately consume them. With the help of this system, not only commercial transactions between producers and consumers, but also all other human interactions will be symbiotic and based on mutual benefit, rather than on greed, exploitation, and enslavement. While the GVE system will help prevent flight of capital, tax evasion, market protectionism, and abusive economic and financial practices in general, it will also help eliminate social conflict, including labor union and management disputes. It will naturally promote lucrative and wise investments of local and global capital, based on Real Value of goods and services, as well as social responsibility and solidarity throughout the world.

In turn, in Chapter 7, we describe the rationale and architecture of a fully automated and interactive center of medical diagnostic and genetic data on-line, which can be implemented at the planetary level. The center uses real-time, health-record data, together with environmental and social data, as data stream, to continuously seed and energize its computational environment. Its technological medical platform, which we have called Zoe, will gather, process, and redistribute, under fully securitized conditions to protect individual privacy, medical knowledge produced through the treatment of millions of patients and genetic information generated by thousands of research institutes and laboratories throughout the world. Much has been done in the past few years in many disconnected projects all over the world in terms of capturing electronic health-records. But, Zoe integrates all of these technological advances in a globally arranged base, or prerequisite electronic environment, and builds up from there.

The full implementation of the project requires the sustained, long-term effort of intercultural teams of researchers in fields such as medicine and pharmacology, genetics, bioinformatics, data management systems, social science, ethics, statistics, health care, intercultural studies, environmental sciences, and so forth, to assist in collecting, evaluating, and organizing medical and genetic data from doctor's offices, hospitals, genetic research institutes and laboratories, medical libraries, national health-record offices, and so on. Once the system is fully implemented, it will greatly advance medical and genetic research, and it will also detect and address outbreaks of epidemics and/or bioterrorist attacks at their incipient stages.

In consonance with the Quantum Relations Principle, Zoe's data and methods of collection and analysis are based not only on the assumptions and practices of Western, allopathic medicine, but also on those of the major schools of so-called alternative medicine. Above all, Zoe is oriented toward preventive instead of "repair" medicine and places healthcare decisions in the hands of the individual patients, offering them the most effective, comprehensive, and integrated treatments.

Finally, in Chapter 8, we propose the implementation of a planetary network of Intercultural Centers of Integrative Knowledge, Technology and Human Development (ICIKs, for short). These centers are designed according to the Quantum Relations Principle and are fully capable of developing and operating the kind of global projects that we have described in this book. The overall mission of the ICIK network is to help re-orient the planet toward a peaceful human mentality and, consequently, toward healthy and prosperous, consensus-based communities, through introducing innovative and effective learning objectives and strategies that will allow them to meet the challenges and take full advantage of the opportunities of globalization.

In line with its mission, the ICIK network will: 1) train ethically and socially responsible political, business, cultural, and civic leaders and entrepreneurs for a global age; 2) promote intercultural research, learning, and dialogue in a local, regional and global context; 3) explore, propose and implement ways of developing new transdisciplinary and transcultural knowledge by integrating the latest research in scientific, humanistic and artistic fields and reorienting such knowledge toward hands-on social and economic problem-solving and creative innovation in the major domains of human activity; and 4) help enhance the contributions of each ICIK's regional communities to the global market by identifying the best ways of translating newly generated knowledge into economic, societal and cultural value. To fulfill its mission, the ICIK network has five categories of activities: intercultural research, education, consultancy, training and exchange.

The ICIK network has two main components: the Global Learning and Research Centers (GLRCs) and the Integrated Data Center (IDCs). These two components are closely linked and continuously coordinate their activities in a mutually reinforcing, symbiotic manner. Each ICIK will choose its own administrative chart and its own research, learning, training and other activities, according to the specific nature and character of the region in which it is located. Nevertheless, its activities must be in keeping with the general mission and objectives of the network and will be closely coordinated with the similar activities of the other ICIK members, in order to avoid duplication and ensure maximum effectiveness.

Once the initial, basic grid is built, other ICIKs or similar centers will emerge as nodes in a worldwide, self-organizing network that will cover the entire planet. It is absolutely essential that these centers should be politically and financially independent: they must not be run by any single government or multinational corporation, although they will cooperate with all public and private organizations, regardless of their ideological or political platforms, as long as such platforms will not be incompatible with the expressed mission and objectives of the ICIK network.

We should emphasize that all of the QR-based technology that we have described here is available and fully implementable today. Indeed, we are already using it actively in many complex applications in fields such as big-data news analysis, geopolitical and macroeconomic models, forensics, and dynamic investigations of holistic transdisciplinary relationships in "in–motion models." Naturally, we prefer to use supercomputers, super-cloud environments and massive parallel databases in order to unlock the power of the Quantum Relations Principle. But we can add substantial benefit by just using a pencil and paper to chart out a certain problem in QR terms. As Part Two of the present book will show, we begin by "writing down" the problem in QR conceptual language and diagrams, before we actually feed it into a supercomputer. We can thus gain superior knowledge and insight about a complex problem even before resorting to a machine to obtain a definitive solution. In our daily interactions with clients from all over the world we demonstrate how consulting work can be successfully structured around the QR methodologies.

We must also emphasize, however, that our methodologies cannot be fully implemented on a massive, global scale without the intelligent, collective effort on the part of the networked global community and a great deal of vision and political will on the part of local, regional and world leaders. It is chiefly for this reason that we propose the implementation of the planetary network of Inter-cultural Centers of Integrative Knowledge, Technology and Human Develop-ment. Indeed, it is our dream and most fervent hope that the 21st century will be known to future generations as the Age of Global Learning, when humanity has finally renounced violent competition, greed, enslavement, and war and has taken the first, but decisive steps, toward planetary wisdom.

Acknowledgements

The authors wish to express their gratitude to several academic and research institutions. First, our thanks go to the Romanian National Authority for Scientific Research, CNCS – UEFISCDI, which made this publication possible through a generous grant (Project number PN-II-ID-PCE-2011-3-0771).

Special thanks are also due to PatentPool GmbH and its head Dr. Heiner Pollert in Munich, Germany, for unwavering support in developing Quantum Relations into a successful technological solution for the 21st Century.

Mihai Spariosu would also like to thank the University of Georgia, at Athens, USA for granting several research leaves, so he could collaborate on this book with Mr. Hardy Schloer in Europe.

We are grateful to a number of friends, colleagues and students who have read this work in its various stages and have offered their kind comments or have helped with it in other important ways: James Anderson, Jason Cornez, Mikhail Epstein, Adela Fofiu, Vlad Jecan, Fateh Maouche, Mirca Pasic, Gheorghe Stefan and Konrad Wienands.

We would also like to thank Sorin Antohi, Chun-chieh Huang and Joern Ruesen for their generous advice and for including our book in their Reflections on (In)Humanity series, to the "Humanity" part of which we hope to bring a contribution; and Ms. Marie-Carolin Vondracek of V&R unipress who has competently and efficiently guided us through the pre-production and production phases of the book.

Last but not least, a very special acknowledgement is due to Philip A. Gagner for his valuable contributions to the QR concept over the years, and to this book in particular, in the area of mathematics and many other areas of science, as detailed in the Brief History of the Quantum Relations Principle, at the end of this volume.

Part One.
The Quantum Relations Principle: Theoretical Foundations

Chapter 1.
What is the Quantum Relations Principle?

The Quantum Relations Principle (QR, for short) is an offshoot of the contemporary philosophy of process, theory of relativity, quantum mechanics, systems theory and cognitive and social science. It brings together, in transdisciplinary fashion, the study of physical phenomena with that of mental and social phenomena, on the assumption of the underlying unity of *all* phenomena. In computer science, QR is a revolutionary, theoretical and practical, measuring tool that allows us to take into account, simultaneously, the experiences of multiple mental observers (or the recorded findings of different observing physical entities) of the same phenomenon, event, or series of events. QR guides us in building dynamic reality models that can measure the interactions of mutual causality among various observers and can make reasonably accurate assumptions about how they may behave and affect each other in the future. In this chapter, as well as in Chapter 2 of Part One, we shall outline the scientific foundations of QR theory in relation to quantum physics and relativity theory. In Chapter 3 we shall describe its basic computational principles and components, whereas in Chapter 4, we shall place QR in a larger philosophical context and discuss its ethical implications.

The QR Scientific View of Reality

QR is based on a complex, multifaceted and multilayered, dynamic view of reality that differs from the linear and reductionist views of mainstream science, based on the assumption of the existence of an objective, physical reality "out there," which is constantly and invariably the same for all times and places. By contrast, QR starts from the premise that what mainstream, reductionist science claims to be solid and objective facts and realities are only the product of selective observation and learning, emotional and cultural filtering, highly subjective belief systems, and the exponential decay of spotty, individual and collective, human memory. Therefore, QR moves away from the Western classical ontological premise of the independent existence of a knowing subject and a knowable object. It postulates that nothing exists independently in our universe and that reality arises primarily not as permanently constituted objects and entities, but as dynamic networks of relations among objects and entities that are in a state of continuous flux.

QR further postulates that everything arises contingent on conditions or events, understood in both a physical and a mental sense. Things do not possess an unchanging, abiding essence. They arise co-dependently, so that physical reality can be described only in terms of relations among objects, entities, and self-organizing systems, nestled within each other and within our universe. In turn, our universe is nestled within larger universes or relational frameworks. QR thus assumes that our universe is an immensely complex web of interrelated, larger and smaller, systems that mutually affect each other as they interact.

By "systems" we mean not just physical, but also mental or intelligent systems that manifest themselves in the form of observations, thoughts, concepts, beliefs, feelings, memories, impressions, opinions – individual or shared. Furthermore, if different mental systems present different versions of apparently similar sequences of physical events, then each description of our universe can be understood only as relative to a particular system and its observer. A system can have a reciprocal relationship with another system, but any description of a state of affairs by one system is interaction-dependent and can only be viewed through the relationships that arise between the observer systems and the observed system at any given moment.

This is not to say, however, that our universe is a "composite view" of the sum of all the observers and their individual dynamic experiences, because a "composite view" would imply yet another dynamic relation of observer and observed. The most we can say is that our universe generates an enormous, perhaps, infinite number of interacting systems embedded within other systems. QR is an account of the ways in which such systems interact, not of the way they are. Indeed, QR assumes that systems are what they do. They can be described not as entities in themselves, but as interdependent networks, engaged in relationships of mutual causality with other such networks.

QR Definition of the Human Mind

The QR definition of our universe applies equally to the human mind. We can therefore describe individual mind or consciousness by the same relational processes that we use to describe physical and other systems. The individual human mind is an interdependent system engaged in relations of mutual causality with other such systems, just like everything that the mind observes or experiences. We, humans define the world in which we live in terms of the relationships we continuously engage in and experience. Physical reality is an individual experience that is not shared by others in its exact same form. Each individual experience is unique in its assembly and compositions, consisting of different relationships and interactions between different systems, embedded within other systems.

There are as many worlds as there are minds to observe, define and react to them. Each individual mind, reacting to its observations is indeed creating a new and unique version of our universe, manifested through the behavior of this mind. Again, this is not to say that the sum total of these individual minds equals a "complete" universe, because such completeness is impossible. However, any observing mind, through its relations with other such minds may build a common, if partial and fluid, network of shared or "collective" consciousness of a particular world, even as each observer-knot of the network maintains its unique accounts, memories, feelings, and experiences of all observing, acting, and reacting in that network.

Minds can be embedded in living things or in machines, or a combination of the two. In combination, the human mind is augmented with the measuring capabilities of a machine, such as a computer, in order to process information packages, which are, in turn, systems embedded in other systems.

Important Caveat: Our definitions are purely functional and do not pretend touching on, let alone solving, the essential, philosophical question of the nature of our mind, or of our brain, for that matter. This essential question can be formulated as follows: does our brain "generate" our mind, or is it simply a vehicle of relaying information from a "higher," perhaps infinite, source? Western mainstream, reductionist and materialist, scientists believe the first alternative to be the case. Western and Eastern sages alike believe the second to be true. For our purpose here, we prefer bracketing this issue, in phenomenological fashion, and describe what can, at present, be observed about the individual human mind and its "ordinary" manifestations in its daily interactions with other individual minds and its surroundings.

So, whenever we speak of the human mind in this book, we mean the ordinary, individual mind as constituted in relation to other such minds and its specific environment. On the other hand, the QR principle hardly excludes the possibility that machine-augmented minds may evolve to a point that they will qualitatively far exceed any observing, conceptualizing and interacting capacity that ordinary human minds may have at present and will be able to act as a quasi-infinite source of information for the latter.

Understanding the Limitations of the Individual Human Mind

QR assumes that properly understanding the present functional limitations (*not* the infinite potentiality) of the individual human mind is a necessary condition for cognitive science in general, and computational science in particular, to make significant progress. Here are a few propositions that future scientific research in these and other fields would do well to consider:

1. No individual, ordinary human mind can claim to know reality in itself and/ or in its totality, because this mind has a limited capacity to observe reality from outside its individual experience or Frame of Reference (FOR)
2. Each individual mind has an incomplete understanding of reality, because unobserved elements of its observations do not become part of its recorded experience as Observer.
3. If we recorded and added up all observed realities by all observers (all minds) inside X (X being a machine, for example), then X may hold a more complex picture of our worlds (or our universe) "outside" our minds. But even with this complex picture, X is still limited by each observer's observing his observations, by the incompleteness of all observed phenomena, and by the known and unknown limitations in quality of the tools of observation, such as the human senses, recording devices, sensors, and our limited ability to comprehend and process such observations on a universal and objective scale.
4. Therefore, all understanding of reality on the part of an individual, ordinary mind or a cluster of such minds, in a single or all instances and in parallel, remains partial and limited.

Figure 1 below represents these functional limitations in a simplified form: How something appears is a matter of perspective, within a temporal-spatial frame of reference. Each view is true. However, truth or reality is the *ideal* sum of many different, known and unknown, perspectives and reference frames at any point in time, and beyond. It is for this reason that QR finds the traditional philosophical distinction between subject and object, as well as the traditional scientific distinction between subjective and objective knowledge, ontologically invalid and epistemologically inoperative.

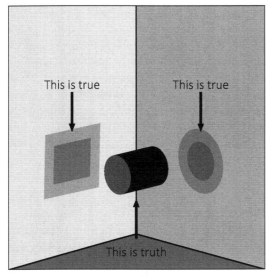

Figure 1: Truth, Perspective, and Time

The QR Model is Probabilistic

Consequently, QR starts from the premise that all observations must be incomplete, non-objective and never totally valid for any measurable amount of time, because constant change in all systems cannot be completely observed, recorded and accounted for in order to "correct" the older versions of reality. Our world and all it generates, including our existence, observations and objects are at best estimations or probabilities of what we claim them to be at any specific moment. Therefore, QR-based technology is a systemic approach of computing reality as best as these limitations will allow and of treating all conclusions it draws from all parallel observations only as dynamic probabilities. These conclusions must never be assumed to be a "true" representation of reality in the traditional scientific sense, but must always be scrutinized and continuously re-evaluated and re-mapped.

Quantum Instances

QR assumes that a human mind functions as a network of associations between quantum instances. A quantum instance is a discrete unit of reality as perceived, imagined or measured (quantified) by a human mind. It also refers to a family of properties that describe one or more mental states. Quantum instances may include individual thoughts, ideas, emotions, sensations, perceptions, dreams, images, or any other category that pertains to a mind's conception or description of physical/mental phenomena. One should, however, not attribute absolute ontological validity to any single mental quantum instance, or even to a cluster of them, because, according to QR, the only viable reality constitutive of a human mind resides in the relations that arise between quantum instances, and not in the quantum instances themselves.

Quantum Relations

The individual human mind forms quantum relations with other systems, building an internalized universe (state-space) composed of these relationships. QR defines a quantum relation as the interaction that arises between an observing system (System S) and the observed system (System O), involving a mutual exchange, transfer, or conversion of small, discrete units of energy or information between the two. Moreover, minds engaged in reciprocal relationships will observe each other, but will, in principle, not "see" the system that they constitute together. They may have complementary ideas, and their definitions of a state of affairs may be strikingly similar, but these may have derived from a different individual life experience and a different underlying system of values and beliefs.

The individual mind becomes filled with relationships (which are all that it can observe) and reacts with output relationships (which is the only way in which it can behave). This is a cardinal mind limitation that invalidates, as we have seen, any claims to objectivity, including scientific ones. Therefore, the only way to qualify and quantify the dynamics and behavior or output of an individual mind is to measure the relationships it forms with other minds and/or other kinds of systems.

For computing purposes, relationships can be qualified and quantified as either "push" or "pull" functions between two or more individual minds, meaning that a mind can receive a relational impulse and then observe and react to it, or it can initiate such relational impulses with another system. Energy or information between systems is transmitted in this way and can be of two kinds: positive or negative.

Quantum Relations and Einstein's Theory of Relativity

QR borrows several principles from Einsteinian physics, applying them to the human mind and cognitive science. In the old Newtonian physics, all movement occurs in absolute space. Once an origin is chosen (e.g., the fixed point [0,0] in Cartesian two-space), all distances can be measured from that point, and all motion occurs against the fixed background of an immutable space, where measures obey the simple rules of arithmetic. Descartes, following Newton (and Aristotle), believed that there was only one true, absolute and fixed viewpoint ("the viewpoint of God"), and everything else moves, and all measurement occurs, in relation to it.

Even in Newtonian physics, however, this is not absolutely true. If I am driving my car next to the railroad tracks while a train is moving in the same direction at the same speed, then the train appears to be motionless. That is, in my frame of reference the train is not moving at all. However, most philosophers and scientists of the earlier paradigm were convinced that there was one metaphysical reality, an ultimate stillness, against which both my car and the train would be seen to be truly moving.

A consequence of the Newtonian model is that objects, for example, have one "true" velocity. They also have a "true" weight, a "true" size, a "true" color, etc. Newton himself was aware that there was no physical evidence that this idea was correct. He adopted it by a method that he called the "first rule of reasoning", or the principle of parsimony, which we now call Ockham's razor. That rule states that given a number of possible explanations, one should select, in the absence of any other evidence, the simplest among them. To Newton, the simplest explanation of motion was that there was an absolute space in which things move.

It turns out, however, that the Newtonian model is not the simplest one. There is no a priori reason to believe that absolute space or time, extending flat to infinity in all directions, is simpler than other models. Alternative models, including Einstein's general relativity seem to have greater simplicity, in the sense of requiring fewer a priori assumptions. In addition, Newton's model cannot account for a significant number of observed behaviors of matter and energy, most of which were unknown in Newton's time.

Einstein's equation describing our universe is this:

$$Rij - 1/2gijR = kTij$$

The single, simple equation replaces the elaborate representation of Newton's forces in space-time. In Einstein's system there is no absolute space. Every movement occurs only with respect to some other object, and the objects themselves generate properties of the space in which they move. Einstein simplified Newton's equation of motion by eliminating any references to absolute position, absolute velocities, and indeed absolute space itself. To achieve this, Einstein discovered that he had to make another simplification – eliminate absolute time as well. There is a bit of artifice going on here. Einstein's equation actually consists of several equations, but expressed in a compact form. Nevertheless, it is also of an astonishing simplicity: The left side of the equation deals with the curvature of space-time, while the right side deals mainly with matter and energy. Another way of looking at this simplicity is that the nature of our universe appears simpler when viewed as a system of four-dimensional vectors in an abstract metric space.

Einstein completely reinvented and made mathematically precise the concept of frame of reference with regard to the motion of objects. He used the concept of transformations of a frame of reference. No longer were reference frames some fixed sets of axis coordinates (Cartesian axes). Rather, space was measured locally, and it changed over time. The frame of reference became rules by which objects interacted. In Einstein's Special Theory of Relativity, the rules themselves could change, but only in very precise and well-defined ways, depending on the relative mass and velocities of the objects within the frame of reference.

Indeed, for Einstein, every object and every observer has its, his, or her own frame of reference. Time may proceed at different rates for different observers. Two observers who measure the same objects and compare their measurements might find out that they have measured different lengths. It is not the case that one observer or the other is right or wrong. Rather, both will have measured the lengths in different frames of reference (moving with respect to each other, for example), and each is perfectly correct within his frame. Einstein's great achievement in the Special Theory of Relativity was to present a method that would predict when this would happen and by how much the measurement would differ. The two observers will measure different lengths if they are in motion relative to each other, v being their relative velocity and c being the speed of light. For example, at velocities such as the 3,000 mph of the space shuttles, the difference in the length of the shuttle to its passengers and to a NASA observer on earth is less than a millimeter.

The concept of frame of reference was quickly adopted in science, mathematics, and philosophy, and with it the notion that an important characteristic of interactions is the way in which their reference frames relate to each other. Indeed, in the 1950s mathematicians developed a set of tools that deal with objects on very abstract levels, portraying their interactions as a set of relationships. This set of tools has been adopted by modern physics and termed "category theory." A major development of modern physics involves the attempt to apply category theory to physical frames of reference. QR adopts this approach, but applies it to mental and social frames of reference as well.

In QR, the important element is not the mathematics of relativity. Rather, it is the idea that objects carry their own frame of reference with them, and that their interactions (making a measurement, for example) are determined by the rules that govern their frames of reference (FORs). QR is an application and extension of the principles of frames of reference or FORs to objects. It provides a description of the elements, structures, and interactions between objects as they evolve over time. The concept of "time," therefore, is central in QR. Most importantly, however, objects in QR need not be physical. They can also be concepts, relationships, or sets of mental, emotional, and social objects. QR attempts to derive, for each such set, the rules that govern the interactions of the objects in the set, as well as the interactions of different sets.

QR, Observation and Heisenberg's Uncertainty Principle

In the early twentieth century, Heisenberg (among others) observed that there is a fundamental problem with measurement. It requires an instrument to make it with, and using this instrument changes whatever is being observed. That is, a measurement always requires that the observer interact with the object measured, and the interaction changes both the observer and the object. Many attempts were made to overcome this difficulty, including calculating the exact effect that an observation could have on the object and the observer (and then calculating the effect of calculating the effect, and so on), in an attempt to quantify and subtract – or calculate exactly – the amount of disturbance and thus to recover the exact state of the object before it was measured. However, this approach failed to produce "corrected" observations that could be explained by physical theories.

Paul Dirac eventually resolved the difficulty by declaring that the observer is part of the system being measured and that there are concrete, physical limits to the degree to which one can treat an observer (including oneself) as an object separate from what is being observed. Both are simply part of one system, so the best one can do is to describe the behavior of the entire system, including the observer. Whenever one tries to extract information exclusively about the object, one introduces uncertainties that render the measurement imprecise. This phenomenon has come to be known as the Heisenberg uncertainty principle.

The uncertainty principle is not a failure of measurement, or of the measuring instruments. It is a fundamental principle of our universe, or at least of the modern quantum mechanical theories about it. This principle states that one simply cannot separate the observer from a system and still obtain precise measurements about the objects being observed. The act of separating the observer's frame of reference from the object's frame of reference introduces the uncertainties.

If we apply the uncertainty principle at all levels of reality, it should be obvious that with regard to any single object (or set of objects), there are as many "realities" as there are observers. Because each observer carries his or her own frame of reference, the results of observations between observers may differ. In addition, any observer cannot be too sure of what he or she has observed. More precisely, there are as many "realities" as there are observers whose frames of reference have different transformational rules. The differences in observation by each observer will differ by the amount determined by the different transformational rules. In relativistic motion, for example, there are as many different "realities" regarding the length of an object as there are observers whose motion relative to that object differs. Of course, these different realities are all related to each other, as seen in Einstein's equation given above.

These physical concepts are most fully developed in the specialized cases of particle physics (the interaction of subatomic particles) and of objects moving in relativistic frames of reference. It is widely assumed that they may apply broadly in the natural sciences, but the Quantum Relations Principle posits that there is demonstrable utility in applying them in the area of cognitive and social sciences as well. It asserts that physical concepts such as frames of reference and processes of measurement can be used in much wider fashion to gain insights into complex processes in general. Even more specifically, the assumption that our universe, at a fundamental level, does not behave like Newtonian mechanical clockwork can bring forth new insights into human cognitive processes, behavior, and personality.

Chapter 2.
Why is Quantum Relations Quantum?

The term "quantum" as used in physics is derived from the Latin word *quantus*, meaning "how many?" It was first used in quantum physics to describe the behavior of systems in which only certain levels were permitted. In classical physics, a particle of light can have any amount of energy, but in quantum physics, only certain levels are allowed. In turn, these levels are highly constrained by the laws governing the atom that emitted the photon. The basic unit of which all other values must be a multiple is called a "quantum." The term "quantized" means that a system can take on only certain values, and those are chosen according to particular rules. As an example, U.S. currency is quantized, and the quantum is the penny. In everyday life, nothing costs a fraction of a penny.

What justifies the use of the name Quantum Relations for the principles and techniques outlined in this book? This is a reasonable question since 'quantum' has a particular meaning in scientific discourse, usually signifying some discrete particle for which further reduction is outside the theory. Therefore, the present chapter explains why this particular name was chosen and describes some of the theoretical underpinnings of the Quantum Relations Principle.

Scientific puzzles discovered in the 19th century led to two revolutionary theories during the early years of the 20th century: quantum mechanics and the theory of relativity. These were more than empirical discoveries; they were both conceptual frameworks within which to remap systems. Relativity greatly sharpened the concept of frames of reference, and quantum mechanics contributed to the development of a well-defined theory of states and state vectors. Furthermore, quantum mechanics was framed in terms of operators (a special method of describing functions) and of probability distributions over function spaces. All of these conceptualizations apply to Quantum Relations.

A Five-Cent Tour of Quantum Mechanics

For our readers who are not mathematically inclined, be assured that nothing in this section involves mathematical manipulations or formulae. Instead, our descriptions are conceptual, to give the flavor of the subject. Quantum mechanics (QM) is a theory of probabilities. It is a model, a conceptual framework used in physics to calculate probabilities. A probability is simply the chance that some event will be observed. For example, the chance that two dice rolled randomly will show 11 is about 55 out of 1000.

A probability relates to future observations (or is used to explain past observations). Probabilities are either calculated or measured. To get to the number 2/9, we could enumerate the different outcomes ({1, 1}, {1, 2} ... {6, 6}) and count the number of each that gave particular sums, or we could roll the two dice a few billion times and mark the number of times that each total appeared. Either method should give pretty much the same answer.

Classical (i. e., non-quantum) mechanics uses ordinary arithmetic, applied to ordinary "real" numbers like 3, 2.7, pi, or seventeen billion. Classical probabilities of events are computable with these numbers, using addition, subtraction, multiplication, and division. If you have enough information about a system, you can calculate any result exactly, and there is nothing in principle, except time and lack of brain cells, to prevent you from gaining a perfect level of information. You can observe, and enumerate or put into some order, all the different states of any system. Single definite states evolve over some time span to other single, definite outcomes.

Quantum mechanics, on the other hand, uses a different kind of number, and a different kind of arithmetic. Additionally, states are not necessarily observable but can still affect outcomes. In QM, all that can be observed are outcomes. Because QM uses a different kind of number, and a different kind of arithmetic, the ordinary way of calculating probabilities will not work. QM has a different method.

QM numbers have two different and independent "ordinary" numbers for every "state" of a system.[1] In many cases, the predictions of the system may not predict a single outcome – there is still only one prediction, as in classical mechanics, but that single prediction may have multiple outcomes, only one of which will be actually observed. For each prediction, every possible outcome will have some probability assigned to it. So, QM, like classical mechanics, makes exact predictions, but these predictions are *not* predictions of the outcome of

---

1 State appears in quotations in this sentence because QM states are not necessarily observable, and therefore are internal theoretical entities that may or may not correspond to anything in the 'real' world.

some process. In classical mechanics, there can be one, and only one, correctly predicted observable. In quantum mechanics, there can be one, and only one, correct set of outcomes, each with one, and only one, probability associated with each outcome. Quantum Mechanics is exact in its predictions of sets, but 'fuzzy' in predictions of any particular outcome. And, worse, one cannot, even in theory, get enough information about the system to make the calculations of a particular outcome exact.

Fundamental to 20th century Quantum Mechanics are two concepts: observables, which are simply quantities capable of being measured; and some function (typically the 'wave function'), which returns the probability (amplitude) of measuring each particular value of the observable. In classical mechanics, one calculates the outcome. In QM, there may be one, two, or even an infinite number of different outcomes, but for each, a probability of occurrence can be calculated. In other words, we do not know whether the answer will be 11, and no amount of knowledge about the system will help us know that, but we can predict that if we keep rolling the dice, eventually 1/18th of the total number of rolls will show an 11.

This idea that only probability (not outcome) is calculable is not unique to quantum mechanics. The way of determining it, however, is. Years of detailed experiments have convinced physicists that simple numbers (3, 2.7, pi, a billion) are not sufficient to explain how things are in the atomic world. However, all the physics known at that time *could* be explained using *pairs* of numbers (such pairs are unfortunately often called complex numbers – in truth there is nothing complex about them).

Since QM numbers are pairs, a new way of doing arithmetic with them had to be devised.[2] Numbers now looked like {3, 2}, rather than simply 3 or simply 2. How do you add {3, 3} to {4, -1}? Multiplication got its own rules too: {3, 0}* {3, 0} gives {9, 0}, but {3, 1}*{3, -1} *also* gives {9, 0}. And, how does one calculate the square root of {6, -3}? QM arithmetic provides an answer, although for our purposes, it is not important what that answer might be.

The QM numbers like {3, 2} are generally called vectors, but vectors can also have more than two elements. A four-element vector would look like {3, 2, 1, 0}. Ordinary (classical) numbers can be thought of as vectors of length 1, *i. e.*, {3} but there is nothing to gain by not merely calling that quantity 3. Classical mechanics also uses vectors, but the elements of these vectors are usually ordinary numbers. But in QM, to repeat, the basic elements are not simple numbers. For-

---

2 "New" is likely the incorrect word here, because mathematicians already knew how to do such manipulations of complex numbers and matrices. In the early part of the 20th century, Heisenberg and Schrödinger independently discovered the value of these artifices to explain and predict experiments in atomic physics.

tunately for physicists, two element numbers appear sufficient to explain physical problems, using somewhat odd definitions of addition, subtraction, multiplication and division. There is no obvious reason for this to be true, but it seems that it is so.

A Five-Cent Tour of Quantum Probabilities

A probability is the chance that something will occur. There is a 20 % chance of probability of rain tomorrow. Earlier we said that QM predicts sets of probabilities. This is not quite true – QM predicts sets of probability amplitudes. Thus, QM is even a little farther away from classical mechanics. There is nothing difficult or particularly strange about probability amplitudes – a probability amplitude is just the square root of a probability, using the arithmetic of QM.

QM also uses probability density. Probability density can be thought of as a map of 'how likely' some measurement is to be found with some particular value (some particular probability amplitude). For example, the probability of finding me on Christmas Eve is more dense around my home, in London, England, than around Melbourne, Australia.

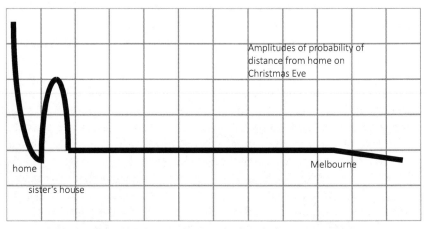

Figure 2: (not to scale) Probability Amplitudes Distribution

Quantum systems, as well as classical ones, have states. Everyone knows what a state is: My car is either in the state of running or the state of not running. A state is also a location in some abstract space – in the car example, the distinct space is comprised of two points, running and not running. Using the example of Christmas Eve, the space could be the surface of the earth, and both Melbourne and my home in London would be regions within space. So, my states could be the group consisting of {near home, near Melbourne}. The probability density would be higher in the 'near home' region than the 'near Melbourne' region. Technically, the probability density is a relationship between quantum states and the (amplitude of a) probability of finding the observable (me) in that state.

Quantum Mechanics, Quantum Relations, and Measurement

When making some physical measurement, the probability of "finding" the observable in some particular state (i.e. having some particular value) is proportional to the square root (in QM arithmetic) of the probability amplitude.[3] That is, QM does not predict the value of any measurement. Instead, it computes the *probability amplitude* of measuring some result. An example would be, as shown above (Figure 2), the set of probabilities of finding me at some distance from my home on Christmas Eve. One fundamental feature of quantum mechanics is that the sum of all the probabilities measured over all the possible values of the observable *must* add up to one. For example, in Figure 2, there could be, say, 59 % probability that I will be near home, and 39 % probability that I'll be near my sister's house, and a 2 % probability that I'll be somewhere else. These total probabilities add up to 1.0.

As mentioned at the beginning of this chapter, Quantum Mechanics is 'quantum' because it posits 'quanta', which are tiny indivisible entities that have states. A photon, for example, is a light quantum, and it has a state that represents its position in physical space. A photon can also be right-handed or left-handed, and or it can be partly in one state and partly in another, although the two states are entirely distinct.[4] Once all the reachable states of a system have been enumerated, plus the probability amplitudes of transitioning from one state to another, QM knows everything possible about that system. So, the idea of state is tightly connected to the idea of quanta. Quanta are elements that cannot be further reduced, within the theory.

In turn, Quantum Relations (QR) also has quanta. A quantum in QR is a "Data Fusion Object" (DFO). Data Fusion Objects can be constructed from other DFOs, and many DFOs can be deconstructed into their components. However, in

---

3 This squaring of the probability amplitude is related to the fact that, in physics, quantum states are located in a complex vector space (a Hilbert space), and complex numbers have two measures ($a + bi$ where I is the square root of -1). Since probabilities are (by definition) real numbers, positive and with only one measure, squaring the probability amplitude and taking its absolute value converts probability density within the Hilbert space into a real number.

4 This ability to be in two distinct states at the same time distinguishes quantum logic from Aristotelian logic, violating the principle of the excluded middle. It is part of the apparent weirdness of quantum arithmetic.

any QR system, there is a point at which DFO's cannot be further deconstructed. These baseline DFOs are the quanta of the particular QR system.

QR also has its own arithmetic. A system comprised of two DFOs will behave in ways that depend on how the properties (i.e. observables and methods) combine. DFOs can be added, subtracted, multiplied, and combined, and, just as in QM, more complicated combinations are possible, such as grouping, encapsulating, and splitting. Still, there comes a point in QR where a DFO cannot be further reduced, and therefore QR is entitled to be called a non-physical quantum theory.

Indeterminacy, Random Variables and Probability in QM and QR

Popular recitations of Quantum Mechanics emphasize *indeterminacy*, including the famous Heisenberg uncertainty principle that we have already mentioned. In QM physics, the quanta are exceedingly small, many times smaller than atoms. When working out problems in QM arithmetic with real world particles, the indeterminacy results from the basic definition of QM multiplication.

The uncertainty principle of quantum physics is worth further explanation, because it reflects a more general principle of certain algebras (in QR terms, arithmetics). As mentioned earlier, QM uses the concept of a probability density. It can be thought of as being denser where things are likely, and less dense where things are unlikely. Generally, the probability density is given by some function, called the *probability density function*, or simply the *pdf*.

Scientists who study QM or the mathematics of probability have devised all sorts of ways of calculating the *pdf* for starting in one particular state of a particular system, then after some time, reaching another particular state. None of these ways is important here, but once the *pdf* is known, you can calculate lots of other interesting things that describe states of the system and their evolution over time.

We shall now break our promise of avoiding mathematical formulae. But, the non-mathematical reader can simply skip this section and move to the next, keeping in mind only that Quantum Relations, like QM, is a probabilistic theory with some basic, irreducible quanta. The consequence of this, from a purely mathematical viewpoint, is that both systems have built-in and unavoidable uncertainty. For QM, the uncertainties are tiny compared to real-world events, and they are unlikely to be observed by anyone other than high-energy physicists. QM uncertainties do not explain psychic phenomena, they do not explain magic, and they do not explain why people forget their house keys. Furthermore, *QR uncertainties are not caused by QM uncertainty and, except for some mathematical parallelism, they are probably not related to it.* But in QR, the calculations of uncertainties have a more dramatic effect, on a much larger scale than in QM, and this results primarily from the inability to precisely specify DFOs and DFO arithmetic that are computable using today's computing machinery. The best we can do is to use powerful supercomputers to create approximations. Having said that, the non-mathematically inclined reader can safely skip to the next section.

For the more adventuresome, let us get right down to the relevant math.[5] Much of this will seem like a review of elementary QM, but certain particular features are worth highlighting. We define some quantity X as a "random variable". This means that an exact value for X is unknown, and perhaps even unknowable. Every time you measure X you can get a different and unpredictable value. X is like your eccentric uncle at dinner – you can never predict what his next unlikely political statement will be.

The random variable X does have some well-defined properties. One is its *probability density function* $f_X(x)$, which is defined such that:

(i) $f_X(x)$ is never negative regardless of x, and

(ii) $p_{a,b} = \int_b^a f_X(x)dx$ is the probability that X has a value somewhere in the interval $[a, b]$, that is, $a \leq x \leq b$. The integral sign simply means that *every possible value* is tested in the region $[a, b]$.

Since we want to determine X empirically, that is, by making measurements, we also require that with probability 1 (i.e., certainty), X has some value (perhaps including zero) *somewhere*. This requirement is stated mathematically as:

(iii) $\int_{-\infty}^{+\infty} f_X(x)dx = 1$

As already stated, there are many ways of deriving a probability density function (*pdf*) for a particular system, whether it be probability theory or QM or QR. For now, we will assume that the *pdf* has already been calculated, that it has some well-defined form, and meets the requirements above. There are some interesting things we can learn from every such *pdf*.

First, we can calculate the *mean* of the *pdf*, which is essentially its average value. We write the mean as $\widehat{X}$ and calculate it in the obvious way, adding all the possible values of X and dividing by the number of such values. Even if the number of possible values is infinite, we can perform this calculation by using an infinite sum, such as:

---

5 We are indebted to Professor Paul J. Nathin for the form of the argument set out below, which can be found in considerably more detail in Chapter 5 of his excellent book, *Dr. Euler's Fabulous Formula*, Princeton University Press, 2006, updated 2011.

$$\int_{-\infty}^{+\infty} x f_X(x) dx$$

It is not obvious, but it is true that we can always make a system in which $\widehat{X}$ is equal to zero, and our new system will be equivalent to the original. To do this, we simply define a new $\widehat{X}$ that is equal to the old one, except we subtract it from X. That is, we work with $- \widehat{X}$, and simply call it $\widehat{X}$. A little thought will show that since X is a random variable, but $\widehat{X}$ is simply a (real) number, nothing has changed with this substitution. Mathematically,

$$\widehat{X - \widehat{X}} = \int_{-\infty}^{+\infty} (x - \widehat{X}) f_X(x) dx$$
$$= \int_{-\infty}^{+\infty} (x) f_X(x) dx - \int_{-\infty}^{+\infty} (\widehat{X}) f_X(x) dx$$
$$= \int_{-\infty}^{+\infty} (x) f_X(x) dx - (\widehat{X}) \int_{-\infty}^{+\infty} f_X(x) dx$$
$$= (\widehat{X} - \widehat{X}) = 0.$$

Remembering that $\widehat{X}$ is the average value, and X is the particular value given by the system at random, it should be obvious that if X does not vary from $\widehat{X}$ by very much very often, then $\widehat{X}$ is a pretty good indicator of the typical value of X. Therefore, it is useful to know when this is true, and we can define a quantity called the *variance* of X as a measure of it. This variance is written as $\sigma_X^2 = (\widehat{X - \widehat{X}})^2$, and it measures the square of the variation of the random variable around its average value. The primary purpose of squaring the variation is to eliminate negative values, which would cancel out positive ones. Combining this result with the definition that makes $X = 0$, and taking the square root, we get

$$\sigma_X = \sqrt{\int_{-\infty}^{+\infty} x^2 f_X(x) dx}$$

This has a great deal to do with quantum mechanics, but also with Quantum Relations. Nothing in the above discussion concerned any physical quantities, just some formal manipulation of a random variable. But, if you think about σ_X for a few minutes, you will see that it represents, in a very real way, the uncertainty that we have in the value of X. A "small" σ_X means that X is pretty certain, not very uncertain, while a large σ_X means that X is all over the map.

Quantum Relations, Fourier Transformations and Uncertainty

In the world of Quantum Relations, it is often difficult to express quantities in some "natural" way, as time series of observable quantities. It may be much easier to express them as frequencies of occurrence, with amplitudes associated with frequencies, or as frequencies and phases. Without getting into the methods of such representations, it will be obvious to the mathematically inclined that this last way, representing a series or a time-amplitude function as frequency and phase, is equivalent to using the Fourier transform of the original series or function.

A function and its Fourier transformation are the same thing, just represented in different ways. A time series function (whether continuous or discrete) is a map of value amplitudes at different times, whereas its Fourier transform can be used as a map of where the *energy* is at different *frequencies,* that is, at different *periodic time intervals.* To say it in a different way, time series show *time value pairs* and the Fourier transform of a time series shows *energy interval pairs.*

Consider a particular function $f(t)$ and its Fourier pair $\mathcal{F}(\omega)$. We could have $f(t)$ represent some location in time and $\mathcal{F}(\omega)$ represent energy at some frequency. Location in time means that "most" of $f(t)$ occurs within some interval of time and "some frequency" means that "most" of $\mathcal{F}(\omega)$ occurs within some interval of frequencies.

The term *energy* needs some definition. We can use the classical mechanical definition quite nicely: The energy of some function $g(x)$ is defined as

$$W = \int_{-\infty}^{\infty} g^2(x)\,dx$$

As mathematician Hermann Weyl pointed out in 1928, referring to the structure of QM, we can show by dividing both sides of the above equation by W that:

$$\int_{-\infty}^{\infty} \frac{g^2(x)}{W}\,dx = 1$$

This is reminiscent of condition (iii) for a random variable. That is, substituting time (t) for x, we can justifiably treat $\dfrac{g^2(t)}{W}$ as a random variable in time.

Therefore, we can compute its mean and variance, and applying the formula given above, $g(x)$ has an uncertainty of

$$\sigma_x = \sqrt{\int_{-\infty}^{+\infty} \frac{g^2(x)}{W}\,dx}$$

For our purposes, and following Weyl, we substitute $f(t)$ for $g(x)$ and get

$$\sigma_t = \sqrt{\int_{-\infty}^{+\infty} \frac{f^2(t)}{W} \, dt}$$

This is quite a remarkable approach. We are treating time itself as a random variable taking values in the interval $[-\infty, \infty]$, and having an average value of zero we derive a probability distribution function of $\frac{g^2(t)}{W}$.

Next, we use Rayleigh's energy formula relating to Fourier transforms, giving

$$W = \frac{1}{2\pi} \sqrt{\int_{-\infty}^{\infty} \omega^2 |\mathcal{F}(\omega)|^2}$$

or

$$\int_{-\infty}^{\infty} \frac{|\mathcal{F}(\omega)|^2}{2\pi W} = 1$$

where for all ω,

$$\frac{|\mathcal{F}(\omega)|^2}{2\pi W} \geq 0$$

What we have shown is that $\frac{|\mathcal{F}(\omega)|^2}{2\pi W}$ looks just like the probability distribution function of a random variable that takes on values in the interval $[-\infty < \omega < \infty]$, and which (consequently) has an average value of zero. Therefore, just as we did with time, we can write

$$\sigma_\omega = \sqrt{\int_{-\infty}^{+\infty} \frac{\omega^2 |\mathcal{F}(\omega)|^2}{2\pi W} \, d\omega}$$

which gives us the uncertainty in frequency.

We have derived an uncertainty in location in time, and we have derived an uncertainty in location in energy. The question that occurs to every student of Heisenberg is: what can we say about the *product* of these, i. e., $\sigma_\omega \sigma_t$?

Mathematically, and again following Weyl, we use the Cauchy-Schwartz inequality: If $h(t)$ and $s(t)$ are two real-valued functions, then assuming the integrals actually exist:

$$\left\{\int_{-\infty}^{\infty} h(t)s(t)dt\right\}^2 \leq \left\{\int_{-\infty}^{\infty} h(t)dt\right\}^2 \left\{\int_{-\infty}^{\infty} s(t)dt\right\}^2$$

For our purposes, we substitute $tf(t)$ for $s(t)$ and for $h(t)$ we substitute $\dfrac{dg}{dt}$. With these substitutions, the above inequality becomes:

$$\left\{\int_{-\infty}^{\infty} tf(t)\frac{dg}{dt}dt\right\}^2 \leq \left\{\int_{-\infty}^{\infty} t^2f^2(t)dt\right\}\left\{\int_{-\infty}^{\infty} \left(\frac{dg}{dt}\right)^2 dt\right\}$$

Where to go from here is not entirely obvious, but since $g(t)\dfrac{dg}{dt} = \dfrac{dg^2(t)/2}{dt}$, then

$$\int_{-\infty}^{\infty} tf(t)\frac{dg}{dt}dt = \int_{-\infty}^{\infty} t\frac{d\left(\frac{g^2(t)}{2}\right)dg}{dt}dt$$

and the integral on the right of the equality can be integrated by parts:

$$\int_{-\infty}^{\infty} t\frac{d\left(\frac{g^2(t)}{2}\right)dg}{dt}dt = \left(\frac{tg^2(t)}{2}\Big|_{-\infty}^{\infty}\right) - \int_{-\infty}^{\infty} \frac{g^2(t)}{2}dt$$

And because g(t) goes to zero much faster than $^1\!/_{\sqrt{t}}$, then

$$\lim_{|t|\to\infty} (t(g^2(t))) = 0$$

which means that

$$\int_{-\infty}^{\infty} t\frac{g(t)}{dt}dt = -\int_{-\infty}^{\infty} \frac{g^2(t)}{2}dt$$

This is almost our classical definition of energy (negated and divided by 2), so that

$$\int_{-\infty}^{\infty} t\frac{g(t)}{dt}dt = -\frac{1}{2}W$$

and this is the right-hand side of the Cauchy-Schwartz inequality. The left-hand side is simple by comparison. By definition,

$$\int_{-\infty}^{\infty} t^2g^2(t)dt = W\sigma_t^2,$$

and we use Raleigh's energy formula in this way:

$$\int_{\infty}^{\infty} \frac{dg^2}{dt} \, dt = \frac{1}{2\pi} \int_{-\infty}^{\infty} \omega^2 |\mathcal{F}(\omega)|^2 dw$$

which is equal to $W\sigma_{\omega}^2$ where σ_{ω} is the *uncertainty* in frequency, that is:

$$\sigma_{\omega} = \sqrt{\int_{-\infty}^{+\infty} \frac{\omega^2 |\mathcal{F}(\omega)|^2}{2\pi W} \, d\omega}$$

And now we can state the Cauchy-Schwartz inequality as an uncertainty principle, relating measurements of time and frequency:

$$\left(-\frac{1}{2} W\right)^2 \leq \left(W\sigma_t^2\right)\left(W\sigma_{\omega}^2\right)$$

which simplifies to

$$\left(\frac{1}{2}\right)^2 \leq \sigma_t^2 \sigma_{\omega}^2$$

so that

$$\frac{1}{2} \geq \sigma_t \sigma_{\omega}$$

This result is important – it means that any attempt to determine future values by modeling past measurements with random variables is subject to a fundamental limitation in its accuracy. And yet, the real world (particularly in QM) appears at some very fundamental level to be best modeled by such methods. Random variables are essential in QR, therefore it is subject to the same limitation, and that the use of time series, whether discrete or not, and models such as the Data Fusion Object (DFO) that are based on time series derived from observations have definite limitations that cannot be overcome by merely adding computing power to them. The result says that the more one focuses on narrowing the precise time slice of a system, the less one knows about what will happen during that time slice.

What Do QM, Relativity, and QR Calculate?

Assume that some set of probability amplitudes has been predicted by a Quantum Mechanics calculation. Nothing is certain about the actual outcome (an observation) except the assigned probabilities. However, once a particular outcome has been observed, no matter how small the probability assigned to that outcome, any subsequent observation will show the same value. This is extremely mysterious – the complicated set of probability amplitudes has suddenly vanished, 'collapsed' into one single and simple, ordinary real number. QM arithmetic assures us that there are no 'hidden variables', that is, it is not as simple as the system having one and only one internal state that we have not observed. Instead, there is a set of values, not just one, and the simple act of observation, by anyone, collapses this set into one, and only one, value.[6]

From a computational point of view, the collapse of the set of probability amplitudes into one simple number makes predictions of complex events much easier. Whenever an observation is made, we substitute the result of the observation for the complicated set of probability amplitudes and continue with the computations. Thus, in Quantum Relations models, probability trees with potentially trillions of leaves equally collapse into a few hundred leafless branches, well within the computing power of today's machinery. We can similarly trim the trees by eliminating the less possible branches (those with lower probability) which, while decreasing the accuracy of our ultimate predictions, makes computation of likely outcomes possible in reasonable lengths of time.

Relativity, on the other hand, is another method used by physics to calculate expected observables. Relativity, as we have seen, is not concerned with probability, but rather with the relationship of physical laws from different frames of reference. A physical principle, according to Galileo Galilei, does not change depending on the observer's frame of reference. In Galileo's book *Dialogue Concerning the Two Chief World Systems* (1632), the physical principle was the relativity of motion – specifically the location of a sack of grain on a moving ship. From the shore, the sack moves. For an observer on the boat, the sack is motionless. However, Galileo argued, there is only one physical principle that must apply in both frames of reference. Similarly, when the ship is moving, a rock dropped from the mast will appear, to the ship's captain, to fall straight down, but, from the shore, it will follow a curved path. And yet, Galileo argued, there is only one physical principle that governs both. One cannot understand any ob-

---

6 This is one of many ways of explaining this mysterious behavior, and roughly follows the so-called Copenhagen interpretation. Others have argued for different interpretations (including the many-worlds approach), but these are not relevant to the present discussion.

servation, then, without taking into account the frame of reference of the observer.

Quantum Relations uses this principle as well. But It does not use (nor does it claim to use) the specific mathematics that governs physical laws such as electromagnetism or gravitation. Fundamentally, QR is the modeling of systems by putting their components into frames of reference and discerning the sets of invariants and symmetries between the observations in different frames and in different combinations of frames. QR is not concerned with Lorenz transformations, or the value of the speed of light, or whether c is a universal speed limit. Nor is it concerned with mass and energy in the physical sense, although it uses similar terms where they help explain a set of relationships or transformations. Rather, it utilizes the methods of relativity in the conception of how different relationships ("laws") can be framed in ways that are independent of particular frames of reference. These relationships are then used to calculate particular results of particular interactions. The theory is successful to the extent that these results (predictions) are successful, i.e., correspond to observed outcomes.

To sum up the discussion in this chapter, Quantum Mechanics is about computing probabilities and using a particular set of machinery to perform these computations. Similarly, Quantum Relations is about probabilities, and not about determined outcomes. QR and QM do not use exactly the same set of mathematical operators, but the mathematical methods behind QR were inspired by the approach of QM. QR is called Quantum Relations because, like Quantum Mechanics, it rejects a privileged frame of reference. Finally, QR does not claim to synthesize either Quantum Mechanics or Relativity Theory. Instead, it borrows mechanical tools (including mathematical ideas) and valuable concepts from both and uses them to calculate real-world results, generally human, sociological, political and economic predications.

Models in QR are valuable to the extent that they are falsifiable, at least in principle. As a relatively simple, common example from the world of finance, consider a set of actors who are modeled in an economic simulation as interacting "wants" and "needs," with limits on some set of internal variables or observables (e.g., wealth). The interactions within some global frame of reference can be calculated, following certain patterns and behaving in certain ways. These ways are often surprising, even to the model designers. When applied to real-world actors, measured values are supplied for the "wants" and "needs," and the simulations can be run on a computer. The results then predict the outcome of the economic transaction. If the result is proved incorrect, then either the supplied values do not represent real values, or the model is incorrect or incomplete.

The benefit of using QR for this type of simulation is that methods of interaction can be described algorithmically (or functionally). These should and can be independent of particular frames of reference. That makes these methods of interaction useful in the same way that other algorithmic or functional processes are useful, as reusable components and building blocks for larger systems. In the next chapter, we shall describe the principles of computation and several specific computational tools derived from the QR principle.

Chapter 3.
The Quantum Relations Principle as Computational Tool

As we have pointed out in Chapter 1, mainstream science measures changes and activities of physical objects and, perhaps, some additional contextual events associated with them through written records, as well as computational static models. But, written records (and current computational models) are flawed and most often incomplete, being compiled through highly filtered and subjective selection of data-fitting information; they are unreliable, and human memories of them can fade quickly.

By contrast, QR measures reality as a complex dynamic web of statistical functions and recorded ecological systems, involving not only traditional physical measurements but also measurements of observational behavior, such as the continuous transformations in observed dynamics of shifting external and internal relations of the observers and their observing experiences. Data and dynamic functions merge into Data Fusion Objects (DOFs) that give answers according to the state of all collective data and computation available at the time we request the analysis. This is a dynamic and computational process that can, in principle, be done by a human brain, but is now much more efficiently done by a machine.

In QR models of human behavior, one fundamental action is that of observation. An observation can be a measurement of a physical quantity or quality (such as temperature). It can also be the result of different kinds of experiments, such as object recognition, or even the formulation of a hypothesis. The process of object recognition is perhaps the simplest example. Assume that we have a machine, called X, which uses a camera to "recognize" a certain face in a crowd. X starts by returning a value of 0, and it continues to do so until it detects a face that matches its template. From that point on, X returns a value of 1. If no face has been recognized, then X continues to return a value of 0. But assume that X is not a perfect machine – sometimes X gives a "false-positive" value (returning a 1 when the face is not in the crowd), and sometimes it gives a "false-negative" value (returning a 0 when the face is actually in the crowd). In a certain sense, one can say that X "observes" the crowd and returns a value with a degree of uncertainty.

Moreover, we are, in turn, observing the machine X. We are hoping, perhaps, that a certain person will appear at the airport, and we have placed X at the gate, checking faces. We check the output of X from time to time. We may be mistaken about our observation of the value returned by X, perhaps because we are too hopeful. Our observations are subjective. Our conclusions are the result of our subjective interpretation of our observations.

The model of which the above is an instance is quite general. It involves a measurement instrument that potentially measures with less than perfect accuracy and an observer of that measurement who (a) perceives the output of the measuring device with less than perfect accuracy; and (b) is subject to misinterpretation (cognitive distortion) of the perception results. None of this requires or involves quantum mechanical kinds of uncertainty, except that it may be modeled by similar assumptions, in the sense that correlations of different measured variables may involve a limiting uncertainty. The uncertainty principle in physics with respect to the energy of a system and time looks like this:

$$DE\ Dt > h/4n$$

The symbol DE stands for the uncertainty in measurement of energy (that is, how sure we are of our measurement), while Dt stands for the uncertainty in the measurement of time. The right side of this inequality is a constant, so it can be rewritten as:

$$DE >\_(h/4n)/Dt$$

This formula has delightful implications: When we take a measurement of a system's energy for some length of time, the more exactly we measure the duration of Dt -> 0, the less exactly we can know the energy (the thing we are measuring). Another way of thinking of this is that there is a deep relationship between energy and time, in which energy can only be defined in terms of time, and this self-referential feature has been given as an explanation of the existence of the uncertainty.

Quantum Relations provides conceptual and computational models for such situations. In fact, QR predicts that because X may return an erroneous value and because we may misinterpret the output of X, we may head to the airport to pick up our passenger at the wrong time. It also makes the stronger claim that we can predict with reasonable accuracy how often we will make this kind of incorrect decision. Of course, other statistical models can do the same thing. What is unique about the QR framework, however, is its ability to take into account the observer's cognitive distortions, subjective prejudices, intentions, and desires. It does this by building a model of human motivation and decision-making, then using that model to interact with the unknown or predicted behavior of X, the sensory device. Such a model is self-referential, and its simplest expression results in limiting uncertainties.

Our QR model uses some of the tools of physics to compute these interactions. Moreover, it uses some of these same tools to predict the processes going on in the brain of the observer. It thus provides a rough model that approximates the computing operations of the ordinary human brain.

QR Objects and Interactors

We shall call "interactors" objects that interact with each other. Objects can interact with other objects in many ways. On a physical level, two marbles can, for example, interact through gravitational attraction, or through an exchange of particles, or through electromagnetic and other physical forces. These interactions are modeled through sets of physical laws, such as the laws of gravitation or of electrodynamics.

Large collections of physical objects can be modeled with the laws of thermodynamics that describe the behavior of hundreds, thousands, or billions of individual interactions through statistical methods. In thermodynamics, or in statistical Quantum Mechanics, one disregards individual interactors and takes into account only the gross properties of their interactions. One trusts that any individual differences in individual interactions will be negligible in comparison to the prediction derived from the knowledge of the physical laws that govern their collective behavior.

When it comes to human behavior, including the behavior of entire societies, we may no longer deal with physical objects and physical laws, but we are still dealing with interactors at multiple levels. One level of interactors may include social entities (people, organizations, countries, etc.). Another level (such as the psychology of an individual person or organization) may include concepts, emotions, memories, habits, preferences, and other mental objects. Observers are also a special case of interactors. When dealing with human behavior, QR is concerned with modeling the "dynamics" that govern the interaction of these social or psychological entities or objects. We posit that this is more than a metaphor – it presents a paradigm for computation, employing a set of tools for studying and predicting the complex relations of such conceptual, behavioral, and personality entities or objects.

QR objects could indifferently be physical, natural, artificial, mental, ideal, etc. For example, a memory or a recording of past events by the brain is such a QR object. However, a memory is not a static object. It changes over time. It may fade, becoming less accessible to the conscious mind, or it may intensify, becoming more vivid. It may be modified by later experience or may change into another memory altogether. Memories interact with other memories, emotions, habits, beliefs, and observations. QR postulates that these changes can be modeled in a well-defined way, using the tools of QR. (Incidentally, nor is a subatomic particle static; subatomic particles viewed within very short time spans may change form, turn into one another and so forth, in accordance with laws governing their interactions, including self-interaction.)

The observation of memory is different from a memory itself. In everyday speech, the "observation" of a memory is the recalling of the memory to the conscious mind. More precisely, it is the way that memory fits into the scheme of interactors associated with conscious self-awareness and behavior. The observation of memory may also change over time, even if the memory itself does not change, because of other factors in the scheme. We call this phenomenon the interpretation of memory objects. The interpretation of memory objects constitutes a significant part of one's individual reality. Therefore, interpretation represents objects of individual reality.

QR Superstructures and Data Fusion Objects (DFOs)

The QR model as applied to human behavior posits certain structures. Again, we do not claim that these structures necessarily have physical existence – merely that they are constructs that, when executed on a computer, allow fairly accurate predictions of behavior. A basic QR concept is that of "superstructure." This simply means that zero, one, or more QR objects can be embedded into a higher-level QR object. The behavior of the higher-level QR object is determined by its own relational rules, plus the behaviors of the QR objects embedded within it.

Thus, a QR superstructure can be viewed as a "space" in which the lower-level objects are embedded. The space has rules that govern the interactions of the objects within it. In turn, the objects have properties that determine how they interact with other objects within the space and with the space itself. We have called such objects Data Fusion Objects, or DFOs. We have chosen this name because the objects "fuse" two different characteristics: properties and methods.

The QR concept of DFOs recalls object-oriented programming languages in design theory, where objects contain their own properties and methods. A property is some value that gets returned to the caller of the object, or is stored internally for later use. In turn, a method is some way of accessing or modifying the information within the object. A simple example of a DFO with properties and methods is a house. A house has various properties, such as its paint-color, the number of windows or floors, location, street address, inhabited or uninhabited, and so forth. A house may have methods as well, such as how to lock or unlock its doors, how to heat or cool it, or generally how to use it as a dwelling.

In the QR model, DFOs are generally interpreted through a frame of reference (FOR). The way this is done is to embed the DFO into a particular FOR, together with the model of the observer. In this case, the frame of reference is also the observer. In most realistic cases, the FOR 'model' itself will be trivial (and it will usually be built into the FOR relational rules). Nevertheless, the fact that the DFO interacts with the observer cannot be disregarded.

More than one DFO may be inserted into a frame of reference. In fact, there may be millions of DFOs embedded within a particular FOR. In turn, the frame of reference itself has both properties and methods, and therefore can equally be regarded as a higher-level DFO. Furthermore, a DFO may be composed of lower-level DFOs (or may be decomposed into them, when the need arises), or it may, in turn, be embedded within higher-level DFOs. In this way, one can build a hierarchy of DFO structures to model any type of problems.

The concept of DFO is overarching, applying to objects, sets of objects, and frames of reference, as well as to various levels of aggregation among them. DFO objects include abstracted objects, such as functors. This flexible hierarchical structure is a fundamental strength of the QR approach. Applying a common concept across all levels and types of "entities" has two important advantages:

1. It allows a common set of analytics and computational tools to be applied throughout the model. At a simple technical level, this means that, in the computer software used to apply the QR model to a particular problem, there are functions to subroutines that can be applied constantly across objects and frames of reference at various levels.

2. It facilitates the development of analytic models that operate on several levels of aggregation (and logical category) simultaneously.

A DFO may be an element in more than one FOR. Each frame of reference can be thought of as a 'view' of that DFO. As noted above, each frame of reference is itself a DFO, so that one can have the following situation:

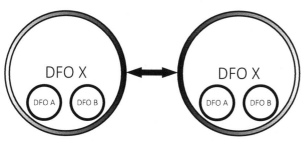

Figure 3: DFO Structures

In this example, the DFO-A on the right and the DFO-A on the left are identical objects, embedded within two different reference-frame DFOs, labeled X and Y. However, X and Y interact with each other, presumably in some higher-level reference frame. X and Y can be thought of as two different "views" of A and B. Their interaction may represent, for example, a way of combining two different statistical analysis of the same data, or even the state of mind of a man who has two different views of the same house – one positive and one less so – and is, therefore, conflicted about buying it or not.

A particular set of "views," each of them being DFOs, may in turn be part of one or more higher-level DFOs. One of the basic rules is that a DFO may communicate only through its higher-level DFO, which imposes an important formal constraint on DFO systems and prevents the DFO model from becoming impossibly complex.

DFO relationships can be expressed in many ways, as demonstrated in the following few examples:

Matching (static) attributes: a1 = a2

"Similar" (static) attributes: f sim (a1,a2) -> 1. The difficulty here is to determine a method to calculate similarity that can be applied in any case. Since ideas of similarity may vary widely, there are also many functions to compute associations and the "neighborhood" of two DFOs' attributes.

Matching instructions: f value (i1) = f value (i2). This is the case when the current main instructions of two DFOs – after being analyzed and evaluated – lead to exactly the same action to take, which is less likely than matching static attributes and yet possible.

"Similar" instructions: f sim (f value (i1), f value (i2)) -> 1. As with similar static attributes, the values of the result action of the two DFOs are compared and associations between them are made.

Interrelated methods: f react (DFO1, DFO2) = 1. If a method of one DFO requires attributes or values of another DFO, they share a strong relationship. This is the case if the DFOs depend on each other, or at least, if one DFO can react on its counterpart (as in a chemical reaction).

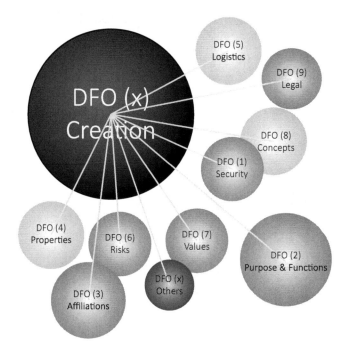

Figure 4: DFO Structures

The DFO structural system of QR is designed to be open-ended to any parallel or higher or lower-level "discovery" of other DFOs. This means that we can choose any known starting point (a DFO in itself) in the physical or mental universe and begin to explore the N parallel or N higher or N lower level systems and their associated dynamics and behaviors. We can record our observations and use statistical tools to predict its further development in the future, given the starting viewpoint of our choosing, or our natural position.

Computing with Data Fusion Objects (DFOs)

A data fusion object (DFO) is an encapsulated computing object. Encapsulated means that the DFO exists in some *environment* where it can read values through the interface between the DFO and that environment, as well as set values in that same interface. Internally, the DFO contains a set of methods that compute its outputs to the interface by combining the interface's input variables with the DFO's stored state, as shown in Figure 5 below:

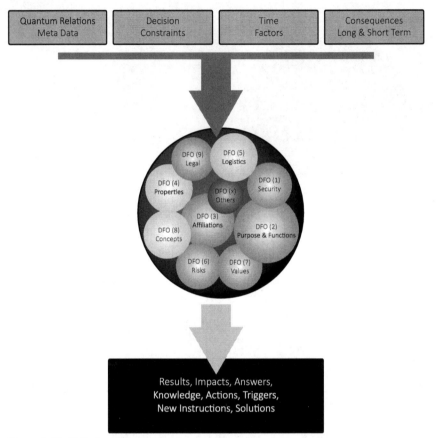

Figure 5: The DFO Computation 1

As already mentioned, DFOs may thus contain other DFOs. The environment of a DFO is itself a DFO, and works in the same way. Therefore, a DFO environment may encapsulate zero DFOs, or one, or any number. From the point of view of the encapsulating DFOs, the internal DFOs are 'atomic', in that the encapsulating DFO cannot inquire into their internal states other than through the DFOs interface. In this way, DFOs are object-oriented, with methods for reading and writing.

A data object within a DFO can be as simple as a constant, or it can be some operator on values read from the environment, or it can be a real-world measurement, or it can be some extremely complicated combination of these things. Whatever its internal structure, the DFO exposes its internal data objects to the environment only through its interface. The interface is a well-defined set of bindings of the DFO's internal states to the exposed "observables" presented to the interface. These observables may be multiply valued and may have multiple methods for access. As an example, a DFO that returns a complex number (a single entity) may, as an alternative, return a pair of real numbers. Where a DFO returns multiple values, but the interface external binding expects a single value, the interface will return the first of the multiple values. Consider the DFO in Figure 6 below:

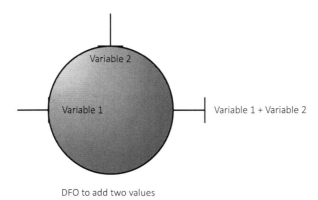

DFO to add two values

Figure 6: DFO Value Computation

Its function is to bind an output to the sum of two inputs. The inputs may be set to some numeric value, or may be some more complicated data type such as a time series. The output will be the sum of the two objects.

A more complicated version of the same sort of DFO is shown in Figure 6a below:

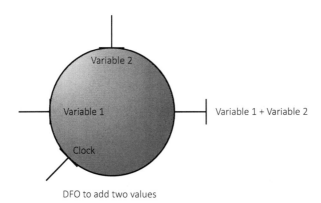

DFO to add two values

Figure 6a. DFO Value Computation

This DFO accesses a different value of its environment (i. e., reads an environmental variable), which is the current time. Therefore, its return value (i. e., the value to which a variable in the environment is bound) is determined by the time that the variable is read.

The difference is very important, primarily because of delayed computation. What is returned by the first example (the simple adder) is not a simple integer (even where the inputs are integers). Instead, an operator on the two inputs gets passed to the environment. That is, if the two inputs are bound to the constants 3 and 4, the output will be the functional object (+ 3 4), which is an instruction to be evaluated later and which, at the time of evaluation, "collapses" into the constant 7.

In the second example, what will be returned is the functional object that can be represented by (+ :at-time <some time value> 3 4). This again is a data object that will "collapse" at some evaluation time to the value 7 (the clock time being entirely irrelevant if the question being asked is the sum of two constant elements). In fact, the interface to the *observer* of the DFO values will remove the at-time modifier and the time value. These will never be evaluated, as they are easily seen to be entirely irrelevant. This removal of computational operators makes the evaluation of DFO systems much faster, and is similar to the optimization functions of many computer languages.

Consider, again, the second DFO example in Figure 6a above, when the two variables are not simply constants, but are time series, and where they exist in an environment in which the sum of the current values of the series can be observed. Such a system is shown in Figure 6b below:

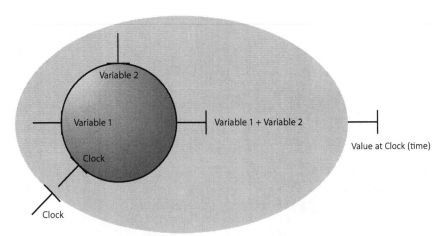

Figure 6b: DFO Computation and Time

The output of such a DFO will be something like:

(+ :at-time <time-value-1> <time-series-1> <time-series-2>), where each of the time series can contain multiple values such as paired date-time and value (assuming the time series is asynchronous).[7] In this example, the values that appear at the right are probably not simply numbers, but are the data type that is requested by the environment of the DFO in which the above DFO is embedded.

This concept of DFO embedding in an environment is one of the critical features of Quantum Relations. Multiple DFOs may be embedded in the environment of a higher-level DFO, and each of these may contain multiple DFOs. In fact, a DFO can be (and often is) recursive, such that it is embedded in itself, and provides its own environment. Such embedding may even be infinite: To make a DFO that provides a never-ending series of consecutive integers, one could (but should not) create a DFO that generates a new copy of itself each clock tick. Each lower-level DFO then updates the value of its caller's return value by incrementing it by one. Nevertheless, this property of recursion is extremely useful in building complex systems.

At some point, the layering of operators upon operators for DFO evaluation must end, and the computations must actually be performed. Before this is done, however, the DFO values (consisting of streams of data whose elements are streams of data) must be themselves evaluated and reduced to a simpler value. This reduction is done at the DFO interface level – the set of boundaries within the higher level DFO where it ends and the environment of its embedded DFO begins. An example follows in Figure 6c below:

---

7 A synchronous time series is a stream of pairs of numbers, where the first number in the pair is a time and the second number is the value of the time series at that time. The first (time) number increases monotonically by fixed increments. An asynchronous time series contains time values that are not necessarily the same distance apart. Asynchronous time series generally (but not always) are sorted so that later times follow earlier ones. It should be noted that the term 'number' in this footnote is somewhat misleading, because both the time and the value can be complicated objects that need evaluation in some higher level if they are to be used as "observables".

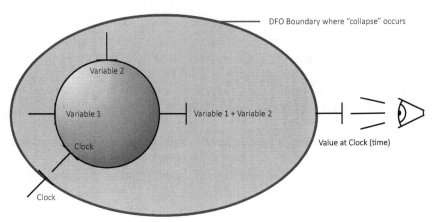

Figure 6c: DFO Structure, Time and Computation

In this example, the pink, most internal, DFO is the same adder used in the previous examples. It takes some values (including a time value) from its environment. The higher-level DFO in which it is embedded does some other computations, but at this level, the values are still operators, and their arguments are not suitable for use as the output of some computational system. The darker blue circle represents an interface that "collapses" the set of operators and arguments into an observable value. In this model, if the inputs are time series and a clock, the output might be a single value of the time-series, which could be a simple number. Everything inside the dark blue boundary is a formal, even symbolic, assembly of operators and arguments. At the boundary, computation (evaluation) is forced and the internal DFO structure is collapsed.

Information may be, and often is, lost at the boundary. In the case where the internal structure contained random variables, the interface must communicate *backwards*, down the chain of embedded DFOs, to ensure that the randomness of the variable in that environment is eliminated and that any future observation of its value will return the same value.

In Quantum Relations, a DFO viewed through a frame of reference (and it can be viewed in no other way) generally returns only particular values according to particular rules. The rules that govern DFO interactions are highly constrained and obey definite patterns. In this sense, QR is a "quantized" version of the everyday human world. The fact that it is quantized means that one can model it on a computer and that one can also perform calculations with it.

In modern physics, a major concern is to create an 'algebra' from a physical theory. An algebra is a set of rules that govern the interaction of elements. For example, ordinary arithmetic has an algebra. The algebra of arithmetic is a set consisting of all the numbers (for the mathematically inclined, it is more precise to say the set of integer numbers) plus two operations, addition and multiplication, plus a few rules of interaction that explain how addition and multi-plication work. Thus the algebra of arithmetic is:

$$\{N,+,*\}\,with\,N=\{0,1,2,3,...\}\ onto\ R$$

The symbol N represents the integers. R is simply all rational numbers, i.e., all the numbers one can get with arithmetic, like $\frac{1}{4}$, or $\frac{1}{2}$, or 110; in fact all the numbers one needs to represent equations with arithmetic operations $(+,*)$ and involving the integers.

In some cases, the objects being manipulated are functions. One can have a set of functions (for example the set of functions f=kx, where k is some constant). Every different value of k generates a different function, and the set of all these functions is the basis of the algebra. These functions can be combined by addition (3x + 4x =7x) or multiplication (3x *2x =7x), which are then two operators on the set of functions. One can create algebras of such sets of functions plus their operators.

Just as there is an algebra of arithmetic or linear functions, one can imagine an algebra of certain types of DFOs. One starts with defined objects and with the operators that one wishes to apply to the DFOs. One can sometimes obtain an algebra over DFOs. Whereas the algebra of arithmetic produces the "field" of rational numbers, the algebra of DFOs produces a different kind of field. We sometimes refer to that field as QR-Space, which generally includes rules that make it a metric space.

Again for the mathematically inclined, functional operators (functors) are developed, and the QR-Space is actually a field of operations over the functors.

QR is "quantum," as we have seen, because it attempts to quantize all types of objects, including concepts, perceptions, memories, behaviors, and so forth. Its objects, however, are generally not numbers but processes, or functions, or "quantitative relations."

Applications of the QR-Space

A metric space is a space in which objects obey certain rules (for example, the concept of unique "distance" between objects is one such well-defined rule). One of the goals of QR is to represent many different kinds of objects – including ideas, emotions, behaviors, etc. – in metric spaces. In some cases, QR uses terms like "mass" or "energy" or "gravity" to model these systems. This does not mean that a system of, say, psychological or personality concepts has actual mass, energy, or gravitational fields. It means only that the objects are modeled by assuming that they behave and interact according to rules similar to the dynamics of physical objects. Because the rules of DFO interactions are part of the frame of reference of the DFO, such rules can be easily written, combined, and tested.

The dynamic modeling idea is a compelling one. Let us take again and consider in some detail the simple, common example, mentioned at the end of Chapter 2, of the dynamics of a system involving several hypothetical stock traders playing the stock market. In Figure 7 below, we derive a point that we shall call the "center of gravity" of these traders regarding certain trades:

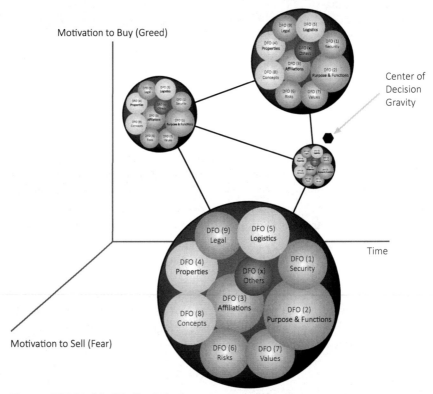

Figure 7: DFO Model of Trading Behavior

We know, or at least speculate, that stock traders are largely motivated by "fear" and "greed," and that these factors influence their preference for a given stock at a given time for some variant degree of intensity and time. We could also use the terms "desire to sell" and "desire to buy," but the terms "fear" and "greed" illustrate more intuitively the presence of different energies as underlying factors of the trader's actions and counter-actions.

We also know that the relative influence of each factor and the degree to which a trader is either attracted or repelled by a given stock change over time. In the picture to the left, various DFO objects are modeled as a function of the "greed" and "fear" of four hypothetical stock traders at specific times when they committed to a buy or a sell trade, each being modeled by one DFO. Thus, let there be:

X = the degree to which a trader prefers the specific stock in the example on the basis of fear;

Y = the degree to which a trader prefers the specific stock on the basis of greed.

One can suppose that for any stock, trader, and time, the trader evaluates his options based on a joint function of a greed motive and a fear motive. Generally this means that:

$$Pi(t) = f, I, t, [x(I,t), y(I,t)]$$

Where

Pi (t) = the preference for or against the stock for Trader i at time t

X(i,t) = the degree to which fear motivates Trader i to buy/sell the stock at time t

Y(i,t) = the degree to which greed motivates Trader i to buy/sell the stock at time t

f,i,t[] = a function that maps the combination of x(i,t) and y(i,t) into a total preference of Trader i to buy/sell the stock at Time t. For example, a simple form of the function might be a weighted composite:

$$f i, t, [x(I,t), y(I,t)] = w i, x X(I,t) + W i, y Y(i,t)$$

Where Wx and Wy are the relative weights of fear and greed in determining Trader i's preferences.

Our time axis in this example represents a typical trading and is repetitive over time. That is, the time axis begins (t=0) at the opening and ends (t=4 p.a.) at the closing of the market every day, with the behavior of the traders averaged over many days. In the diagram, the size of the DFO indicates its "mass," which might be the total trading dollars available to the trader on average.

The position on the time axis indicates the time at which the particular trader is likely to buy or sell a particular stock. The trading behavior of one trader greatly influences the behaviors of the others, so that the actual model is dynamic, not static as shown here. In our diagram (left), there is a "critical point," which is the captured center of gravity of this particular situation. That point represents a "balancing" of the behaviors of the individual traders. In the neighborhood of this point we can expect certain behaviors in the simple market model to start changing.

The averaging element is worth describing in more detail. The value of a trader along the axis "fear" is actually shorthand for the trader's observed behavior over many days. For example, our trader may pass up opportunities that involve "objectively" perceived risk, even as other traders take advantage of them. Here the model postulates that the trader's behavior is driven by fear, but if one introduces any other factor, such as laziness or inability to marshal resources, the model will still work.

If necessary, the property "fear" can be decomposed further, in order to create a more complex model. But that does not substantially affect the model's viewpoint, except that the paths of the DFOs through the space will change over time. The decomposition of "fear" that is applied to the stock trader scenario might be applied to other scenarios as well and may give better results in those systems, or vice-versa. Finally, the entire DFO model may be viewed through one or more frames of reference, and these may be combined to build a model of the mental states and dynamic behaviors of an entire financial market, as shown in Figure 8 below.

In this model, time has been turned sideways to emphasize price fluctuations, while selling transactions are indicated in red and buying transactions, in green. Price in the hypothetical market is determined by the buy-sell transactions and is overlaid to show price changes over time. The selling transactions drive the prices upward, and the buying transactions (which on the graph occur later) drive the price down. The transactions again have a size (which may be thought of as their "mass"), and centers of gravity may be computed. In this example, the centers of gravity correspond to locations on the price curve where changes in kind occur (a rapidly rising market flattens, then begins to drop rapidly). The center of gravity points are critical points in some space that generates the price curve.

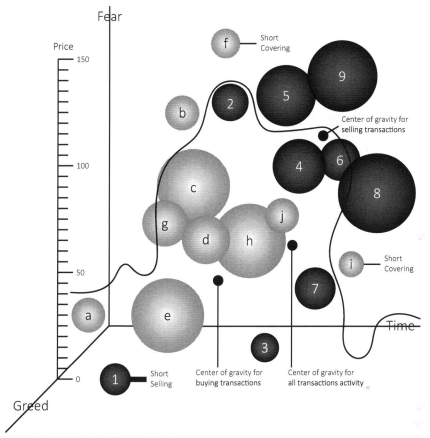

Figure 8: DFO Model of Financial Markets

Time in QR-Space

QR systems evolve over time. This means that the functions governing the interaction of sub-objects (i. e., some FOR rule) depend on a parameter "t."

For example, one can imagine a frame of reference or FOR containing two DFOs, called CLOCK 1 and CLOCK2. These two DFOs as well as the FOR itself have a method, called "tell-time." When the FOR invokes the tell-time method for itself, it gets answer 23. When it invokes it for CLOCK1, it still gets answer 23. But when it invokes it for CLOCK2, it may get answer 27.

In this example, CLOCK1 and CLOCK2 are experiencing time at a different rate. QR has developed some basic (default) principles of time in QR systems, as follows: 1) Time is unique to systems and observers (i. e. the rate at which things happen is dependent on the frame of reference); 2) Time is measured through increments of change within systems. That is, time is discreet, not continuous, although in some cases it can be modeled through differential equations, which assume continuity. 3) Time advance (that is, the distance between 'ticks') can decline to *zero* in a system, without ending the system's existence.

A DFO concerned with time can always be asked what its (local) time is, but the answer may be different for different FORs. In this system, all DFO time is local. However, two DFOs that interact through a properly constructed FOR will never need to worry that their times do not match. They will perceive the passage of intervals according to their interaction with the FOR, and for each DFO the time will always be "the present."

DFOs and Self-Organizing Structures

A self-organizing structure is a structure in which (a) the elements obey certain rules; and (b) the rules modify the structure over time in such a way that certain features are emphasized. Figures 9, 9a and 9b below represent simple examples of self-organizing structures with two- and three-dimensional arrays of numbers:

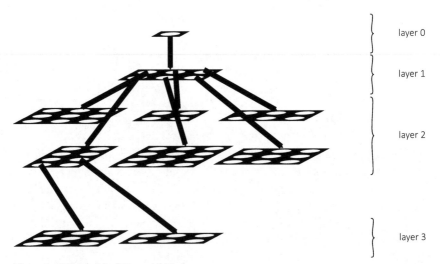

layer 0

layer 1

layer 2

layer 3

Figure 9: DFO Model of Financial Markets

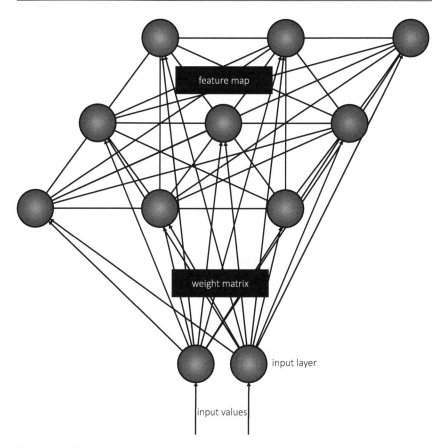

Figure 9a: Self-Organizing Structure 2

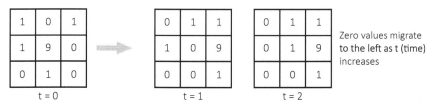

Zero values migrate to the left as t (time) increases

Figure 9b: Self-Organizing Structure 3

In Figure 9b, for instance, the organizing rule for each element is: "If I am equal to zero, and not already on the left margin, then switch me with the element on my immediate left." This rule moves all the Zero values of the array to the left. It may be a useful rule if, for example, the array is very large and one wants to organize the Zero value towards a goal, where it is useful to have as many Zero values to the left as possible.

Many more useful examples can be found in the literature of self-organizing maps. Figure 10 below implies a general model of DFO processing, based on the same self-organizing principle:

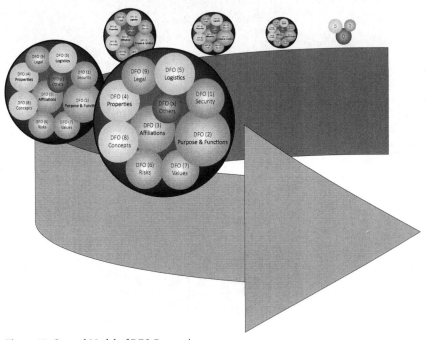

Figure 10: General Model of DFO Processing

It is often useful in QR systems to make the FOR rules self-organizing with respect to their DFO elements. In this case, the passage of time will itself drive the DFO into organized behavior, and this can be used, in some cases, to find patterns within the DFO structure that were previously not recognized. In such cases, the FOR must have appropriate algorithms of pattern recognition built into it, exactly as with any other self-organizing, map-based system.

One difference is that in QR the pattern-recognition systems themselves can be DFO-based. This gives rise to a natural parallelism, because the sub-DFO objects can be shared by multiple FORs that can run on parallel computing hardware or even computers in different locations.

Adaptive Processing in the DFO Model

As described earlier, the manner in which DFOs communicate with each other within a FOR is determined by the rules built into the FOR. These rules need not be static. Indeed, a DFO can pass on to the FOR a suggested new rule for further processing.

A FOR can, moreover, create new DFO objects within itself. It can create, for example, copies of any of its member DFOs, and it can also modify the new copies in any way permitted by the DFO rule sets. One of the methods of a DFO can be a request for a new "strategy" to be used by the FOR in which the DFO is embedded. The FOR can request from its own higher-level FOR that a copy of itself be made, with the new strategy replacing one of its own rules. The requesting FOR can then be either preserved or destroyed. Since the newly created FOR is an exact copy of the requesting FOR, except for the requested changes, the general rule is to destroy the old FOR.

This QR feature gives rise to a flexible and adaptive processing model in which DFOs can interact, as well as be created or destroyed. Their interaction may in turn produce new DFOs. In this respect, the QR model is similar, again, to sub-atomic physics, where particles can also interact and produce other particles, or be annihilated.

A DFO can request its own annihilation by its FOR when its processing tasks are complete (or determined to be futile). For example, a DFO that is used for object recognition would request its own annihilation when it has passed its best guess up to its FOR, the FOR has extinguished all the rules that refer to the DFO so that there is no further need for it.

On a conceptual level, a DFO may be seen as attempting to resolve a particular query, such as whether a given set of inputs matches some template for object recognition. Competing DFOs within the same FOR may be attempting the same task, and the FOR may determine that it is satisfied, simply extinguishing the sub-DFOs. DFO results may be passed on as probabilities, and when a probability passed upward during an ongoing computation reaches a given threshold (such as 90 %), the FOR may be sufficiently satisfied to stop the other sub-DFO objects, releasing resources for other computations.

A DFO is created by its FOR (a higher level DFO) for the purpose of solving tasks. These tasks can be seconds or years or even permanent. This is a process similar to a biologic or ecologic event. First, the system creates different instruction particles with different tasks to do. This becomes the DNA of a DFO. Each particle has "Properties" and "Methods" build into its "native information stem." A new DFO is thus created. Figure 11 below shows an example of how a Security DFO is created, instructed, destroyed, and replaced by a new Security DFO:

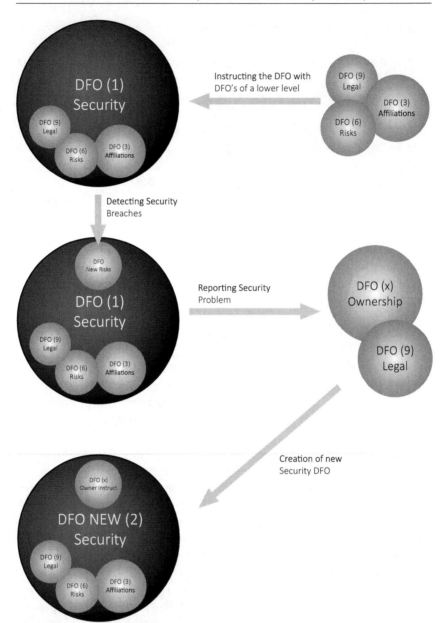

Figure 11. Security DFO Processing in Runtime.

In this particular example, a Security DFO gets loaded with instructions of legality, risk and affiliations. The DFO carries out its instructions and detects new risks that it reports to the Ownership and Legal DFOs. These, then, destroy the old Security DFO and replace it with a new one.

In turn, Figure 12 below presents a general model of creating and instructing a new DFO:

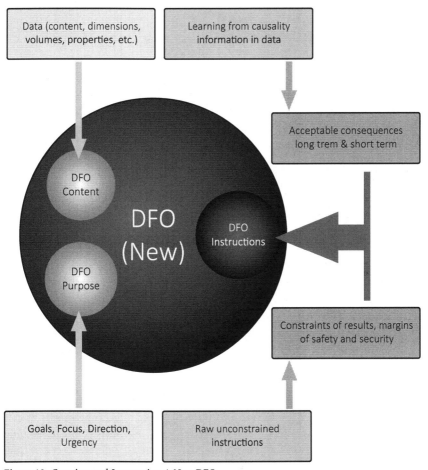

Figure 12: Creating and Instructing A New DFO

The new DFO is created by inserting its three essential particles: purpose, instruction and content; these are in turn DFOs. The Instructions DFO shows the provenance and modality of creation of the new DFO by other DFOs. Each of the hexagon-shaped labels is a filter, representing a higher level DFO, with the purpose of creating the new DFO.

DFO Particles

A DFO can have only a few instructional particles or millions of them, depending on its purpose. As mentioned in the previous paragraph, there are at least three types of elementary particles needed to create a functional DFO:

The Purpose Particle x(u)
The Content Particle x(s), and
The Instruction Particle x(i)

Every particle, regardless of which type, consists of these three different sections of information. In turn, the Genetic Particles Code (gpc) describes each single particle in a unique way and allows for a classification of all particles. Also each type of particle has a certain number of methods that are specific to the type. At least each type of particle has a unique set of attributes.

A FOR can create different types of DFOs depending on the specific tasks such DFO must fulfill for the FOR in the future. Here are some examples:

The Logistic DFO, DFO(lot), which brings all functionality to the system that involves the flow-modeling of recourses among all utilizer members of a system. This guarantees that the logistic prerequisites are in place to reach their targets on time and budget.

The Transaction DFO, DFO(tra), which carries out the execution of the action after the system has reached a decision. It will use only the methods guided by the DFO(lot) of the same synthesis to manage all its transactional task resources.

The Report DFO, DFO(rep), which manages all the reporting functions between the Host Environment (HE) and the utilizers authorized to receive them.

The Communication DFO, DFO(com), which facilitates communication, presentation, style, timing and uses the presence of a DFO(lot) to choose the proper logistics to carry out the communication task.

The Security DFO, DFOsec, which handles the entire security and defensive system of a respective FOR within the host environment HE. There are two types of DFOsec: DFOsec in detect mode DFO(sec.d) and the DFOsec in active mode DFO(sec.a)

The Safeguard DFO, DFO (saf), which handles the systems' safeguard features. It prevents a system from executing processes that would be in contrast to the baselines, ethics and principles of the respective frame of reference (FOR) within the host environment (HE).

The Organization DFO, DFO(opt), which handles the systems optimization features. It optimizes the system beyond the baselines, ethics and principles to implement a host operator trend (HOT) and to focus the host environment (HE).

This enforces master instruction in situations where a specific type of outcome of a process is desired.

The Possessions DFO, DFO(pos), which invokes and guides possessions of expressible value. Its Stuff Particles x(s) describe things in inventory, accounting data (positive and negative) to property of any kind, including but not limited to intellectual property and assumed or desired future assets of any kind (the Greed DFO).

The Logging DFO, DFO(log), which records every event and all its results to the cumulative long-term memory of the host environment (HE). It updates all relationships in the Management System (AbSys) for real-time management dynamics.

DFOs and Locality

DFOs are, from their own perspective, local: in other words, they are self-contained objects that communicate with a higher-level process only through very structured mechanisms. However, the ability to have the same DFO in more than one FOR makes the model globally non-local. Information may be passed through a DFO in both directions, so that two different FOR objects can potentially communicate.

It is entirely possible for a DFO to be operating on one piece of computing hardware (including the "wetware" of some portions of the brain) and to be passing messages to its FOR, which is operating on an entirely different piece of computing hardware. The DFO model accommodates parallel computing technologies across the human-machine barrier. QR is therefore well suited to investigate man and machine within a common environment.

For DFO structures that have been metricized, as described above, the objects are distributed in some space. They are subject to and obey well-defined rules, presumably rules that preserve both the metric nature of the space and the desired space properties. Their interactions can be computed either as static functions or using some dynamic set with local clocks that advance in time.

Again for the mathematically inclined, the functions that a DFO can execute may be viewed as operators on the space-time within it. A cover can be defined which is the set of such functions, and it is often useful to speak of the properties of that cover. Indeed, the behaviors of the DFO itself may be, in appropriate cases, determined by the function space that can be so derived.

Practical Consequences and Implementation of the QR Systems

To sum up, here are the main points resulting from the discussion in this chapter:

1. QR systems are conceptual objects constructed through models of observed systems. These models may be more or less detailed, and they may contain objects that are statistical only, that is, whose behavior is simple and aggregated, or an estimated function of some sub-behavior. It is sometimes possible to decompose such aggregate DFOs into a more precise model, or to aggregate a system of separate DFOs into an aggregate behavior.
2. QR objects and systems are structured and may, in appropriate cases, be viewed as distributed in some space-time.
3. Although DFOs are local from their own viewpoint, QR systems are non-local in time or in their position in space-time. Nevertheless, they are subject to stringent transformational rules imposed by their frame of reference (FOR).
4. QR systems can be distributed and can take advantage of parallelism.
5. QR objects (DFOs) are building blocks or models of observed reality

Given these features, what are the technical advantages of implementing the DFO/FOR model?

For one, this model is self-adaptable and will automatically search for the best method and the shortest path to accomplish its goal. The Quantum Relations Principle assumes that (a) all intelligent systems in motion tend towards ultimate efficiency; and (b) ultimate efficiency results in the equilibrium of all natural, free-flowing forces. Therefore, Quantum Relations technology is designed to replicate and energize this process, both conceptually and systemically, inside an intelligent machine. In this regard, Quantum Relations can be seen as *the operating system of general intelligence par excellence.*

The system can also easily change metrics since the metric distance between DFO and FOR is stored as a property of the FOR. A metric change is the equivalent of asking for a different interpretation of the underlying data. Because metric distances between DFOs and FORs are implemented in a hierarchical fashion, one can easily change perspective on an entire data set. As DFOs implement class inheritance, such changes might ripple down through various levels of sub-DFOs, triggering re-computation of intermediate results in a controlled and natural fashion.

By the same token, the DFO/FOR model is capable of self-organization, because data and functions are implemented as sets of hierarchical objects. For example, if the metric of a FOR is differentiable over the set, data in that set can be concentrated by finding the minimum of the differential, just as in the case of physical models. A FOR containing many DFO structures can also contain rules for the creation of new DFOs; for the interaction between its DFOs; and for the calculation of functions between smaller DFOs, including the creation of new objects that embody certain relationships between these smaller DFOs.

DFOs and FORs are based on a complex network of parallel relationships. These relationships can be expressed as positive (attraction) or negative (repulsion). The interaction between two DFOs can include changing properties of the mental particles themselves, much as, in a physical system, an attraction is a function of space that operates to change the position of objects. A reasonable FOR can implement certain rules of symmetry and conservation among its DFO objects. In this way, the model uses mathematical and physical methods to create a framework within which large-scale computations can be performed.

DFOs and FORs provide a natural model for general parallel computation. Since DFOs and FORs are discrete objects, they can be implemented on multiple processor systems, and calculations can be performed in parallel. The DFO/FOR model is not bound to the theoretical requirement that either the metric or other functions provided by a frame of reference be Turing computable functions. Any function that can take one or more data structures as arguments can be implemented within the DFO/FOR model. Therefore, this model can also provide conceptual methods for implementing quantum computing, as soon as hardware becomes available for such applications. On the other hand, it can equally simulate non-local functions, such as are found in Quantum Mechanics, and implement them on a Turing-Church type of processor (digital computer).

Furthermore, the DFO/FOR model is both modular and extensible. This means that a set of computations on one data set can be transformed into another data set and used by the second data set to define a set of new functions, translating the preceding FOR into the new one. In addition, a FOR can contain rules for logical inference and deduction that operate on its component DFO objects. The fact that FORs are also considered DFOs for higher-level frames allows lower-level frames to define data properties. DFOs could equally be used to pose queries on other DFO frames. This means that both the query DFO and the answer DFO would exist within the same FOR structure until a computation would achieve the goal of relating them. In this manner, the DFO/FOR model can implement functional and rule-based languages such as Prolog in order to solve real-world and hypothetical problems. It can also translate and incorporate any software program or computer language into its database, thus solving the currently intractable problem of systemic compatibility and interchangeability in computer programming.

The DFO/FOR model is compact and adaptable, expressly designed to handle extremely large quantities of data, on the scale of gigabyte and terabyte sets, and to provide methods for manipulating such "big-data" and "meta-data" through parallel processing systems. Under the Quantum Relation Principle, data and software (instructions) become fully unified and act together as a single object.

The more data, the more complete becomes the replication of the actual world and its all-encompassing behavior inside the machine, and the better the system provides solutions to problems. This is a unique, revolutionary feature of QR-systems.

DFO/FOR structures can be compiled, i. e., translated from a symbolic form into a compact set of machine instructions and can also run continuous restrictions on data in order to prevent database errors. The DFO/FOR model can handle data storage, recuperation, and processing with great flexibility and practically no data loss. It assumes that no piece of information or knowledge from its database can ever become obsolete, because it may always turn out to be relevant in a different DFO/FOR configuration, or coherence, or correlation between data sets.

Finally, DFO/FOR structures can build comprehensive real-world models for handling data, together with very complex sociological and psychological information, based on observed relationships and behavior. These models can analyze human cognition, personality, and social behavior on a very large scale. It is important to realize that we have already successfully implemented such models in finance, intelligence, and big-data news analysis applications. Although much more remains to be done, we anticipate that, through continuous development, the QR models will be useful in analyzing the most complex aspects of human behavior on a global scale and in bringing micro and macro sociological observations and predictions to an acceptable level of accuracy. Other applications will be implemented in the near future, such as QR systems for early detection and possible prevention of ecological disasters, global health hazards and epidemics, wars and political terrorism, stock market crashes, severe socioeconomic crises and conflicts, or simply irresponsible and dangerous human behavior.

Down the road, QR applications could include the proactive detection of unsound or irresponsible governing on all levels, the facilitation of global research and learning systems, and elaboration and testing of new scientific hypotheses and new technologies in a wide variety of fields. In Part Two of the present book, we shall describe the electronic architectures of several of these applications in some detail. Before that, however, in the next and last chapter of Part One we shall outline the general philosophical and ethical implications of QR, which are equally embedded in our technological applications.

Chapter 4.
Quantum Relations Theory: Philosophical Underpinnings and Ethical Implications

As we have seen, Quantum Relations starts from the basic insights of quantum physics and the theory of relativity. At the same time, however, it argues that these insights should apply not only to the physical world, but to the human world as well. In this sense, QR is both a critique and an extension of the principles of quantum mechanics and special relativity. Although these theories recognize that the presence of the observer modifies the nature of the phenomena observed, they do not act on this recognition in a radical and consistent manner. Whereas quantum mechanics and general relativity provide reasonably accurate descriptions of how our physical world works, they leave out a crucial component that is the basis of any scientific observation and insight: the human mind. Although some physicists have suggested that consciousness or awareness is a quantum process (Bohm 1990; Stapp 1993; 2011), no successful attempt has so far been made in the physical sciences, let alone computer science, to integrate the mind and mental processes into our modeling of the physical world. QR seeks to fill this gap.

Mainstream reductionist science has mostly seen itself as an objective and impersonal search into the nature of reality. QR emphasizes the need to shift scientific thinking away from the quest for objective Truth towards the recognition that all scientific data are observer-dependent and that all approaches to reality, including scientific ones, are influenced by subjective experience. QR departs from Cartesian dualism of subject and object, or mind and matter, by placing mental and physical events or processes on the same experiential continuum. Consequently, it describes mind or "consciousness" by the same relational processes that contemporary theoretical physicists use to describe physical and other systems.

QR also subscribes to the main assumption of general systems theory that our universe is a web of interrelated systems that mutually affect each other when they interact. Prominent general systems theorists such as Magoroh Maruyama, Ervin Laszlo and Fritjof Capra, among many others, challenge the reductionist model as being a mechanistic, linear paradigm. They define a system not as a static object, but as a dynamic flow of interactions or pattern of events, whose existence and behavior derive not from the nature of its components, but from their manner of organization.

Like general systems theory, QR assumes that a system organizes and maintains itself by exchanging matter, energy, and information with its environment, that is, with other systems. Thus, systems frame and are framed by other systems in a natural, hierarchical order. But in QR, just as in general systems theory, the term "hierarchical" should not be understood in linear fashion, as disciplinary order, but only as larger or more inclusive levels of self-organization. Systems, subsystems, and suprasystems are nested within one another, rather like sets of Chinese boxes. But, unlike such boxes, they constantly communicate and interact.

Although QR theory dispenses with the dichotomy of subject and object, it does not reject altogether the concepts of organism and environment, as they have been handed down by natural science. Instead, like general systems theory, it acknowledges the inherent unity of organism and environment, as well as body and mind, through the complex networks of relations and interactions that emerge among their components. Thus, QR conceives of experience "as linking organism and environment in a continuous chain of events, from which we cannot, without arbitrariness, abstract an entity called 'organism' and another called 'environment'. The organism is continuous with its environment, and its experience refers to a series of transactions constituting the organism-environment continuum" (Laszlo 1969, p. 21).

On the other hand, the brain that the neurologist observes cannot be equated, as reductionists do, with the experience of the mind to which its synaptic activity corresponds, because brain and mind are not on the same continuum. For this reason, QR adopts Laszlo's notion of biperspectivism, based on Niels Bohr's principle of complementarity (in modern physics), which can provide a more accurate description of the correlation between mind and matter than the one that reductionist cognitive science provides. As Laszlo points out, "whether a system is physical or mental depends on the viewpoint of observation. The operation of passing from the one to the other viewpoint permits the alternate inspection of the complementary (but not simultaneously appearing) aspects" (Laszlo 1973, 171).

Matter and mind, body and consciousness are not ultimate realities. They have no rigid boundaries, being simply different conceptualizations to bring order into experience. As a conceptualization or metaphor, biperspectivism allows us to deal with both sides of experience without essentializing either of them. As Laszlo suggests, "complex open systems such as human organisms have an exterior and interior dimension, so that mind is co-extensive with the physical universe" (Laszlo 1973, p. 293).

QR also acknowledges that the traditional notion of causality, defined as a linear, local and physical relation, is inadequate for describing the complementarity of mind and body. Just like general systems theory, QR replaces the notion of linear causality with the nonlinear and nonlocal concept of mutual causality, providing a much more complex, qualitative account of the reciprocal relations among the systemic networks that our minds and bodies constitute.

Unidirectional causal thinking presupposes an essentialist dichotomy between subject and object that depends on static notions of identity, substance, and attributes (Maruyama 1974). By contrast, mutual causality is two-directional, presupposing a dual relation of interdependence and reciprocity between cause and effect. It implies viewing reality as a web of dynamic relations, where substances and identities continually shape and reshape each other and where subject and object are dissolved in the ceaseless flow of events and experiences. This view of reality privileges neither the material nor the spiritual, neither individual nor collective consciousness, neither human agency nor physical nature or environment. Instead, it regards all of them as complementary parts in an experiential continuum, ceaselessly engaging in causally reciprocal, feedback loops.

In turn, the concept of feedback loops is as valuable in QR as it is in general systems theory. This concept was first used in cybernetics to describe feedback mechanisms in various closed-systems or machines. General systems theory then emancipated it from its mechanistic implications, applying it to open, self-organizing systems, including individual and social organisms. Indeed, causally reciprocal loops can be seen as operating across a large number of fields, from the natural to the social sciences to the humanities. They are also present in evolutionary and environmental theory, where they can account for evolutionary complexity and change in more elegant and plausible fashion than unsophisticated, reductionist notions of linear causality.

Quantum Relations also takes into account the interplay of "negative" and "positive" feedbacks in the complex relationship between human individuals, social organisms and their environment. "Negative" feedback, or what Maruyama (in Milsum 1968) and Laszlo (1973) call Cybernetics I, is a self-stabilizing activity that allows individuals to live in a world that they have constructed through their past experiences. By negative feedback loops, individuals generate conditions in their environment that confirm and correspond to already existing cognitive patterns, through which they in turn perceive this environment. Thus, negative feedback is a form of existential projection. The system, whether it is an individual or a community, projects on its environment the constructs that best match its perceptions.

Changing conditions, however, can lead to a mismatch between a system's perceptions and its constructs, triggering "positive" feedback loops, or Cybernetics II (Maruyama 1963, Laszlo 1973). Since its experience no longer conforms to its preconceptions, the system develops new constructs in order to change and refine its previous map of the world. This remapping of knowledge is what we have called "learning" in this book. Just as a natural biological system self-organizes and stabilizes itself in order to adapt to changing conditions, so does a cognitive system, in order to make sense of its new world. The feedback process by which the cognitive system generates meaning both alters the environment, through the system's projection of its constructs, and modifies the system itself, as it remaps this environment (cf. Laszlo 1973, 128–31).

For Quantum Relations theory, just as for general systems theory, learning means not merely becoming acquainted with the characteristics and the pattern of organization of an already existing system. Instead, it involves a fundamental reorganization of the system, in which new assemblies occur, different feedback loops arise, and alternate pathways emerge. In other words, we develop a four-dimensional, or dynamic view of systems, in which we interact and learn through the partial uncovering of past, present and potential future causality, compare the potential outcomes of different responses, and then respond dynamically by choosing the pathway that most closely represents the desired outcome. In the end, both the world and its explorer become transformed through learning. It is in this sense that we are using, in Chapter 8 of the present book, the notion of local-global learning environments, through which we can learn how to reorient ourselves within a planetary reference frame. The planetary frame involves different rules and organizing principles than those within local or regional reference frames while reintegrating and remapping the latter as well. As we pointed out in the Preface, such global learning processes involve an extensive reorganization of all living systems on this planet, including ourselves, and can hardly be achieved overnight.

The notion of positive feedback as learning also implies recognition of the potential creative role of cognitive or social crises. When obsolete modes of interpretation become dysfunctional, confusion and disorientation set in, because nothing seems to work any more "as it used to" – a complaint that is nowadays heard only too often from all kinds of mainstream practitioners and experts, called on to puzzle out the emerging global economies and geopolitical configurations. This disorientation can then trigger a generalized state of anxiety and distress, as indeed many of us are experiencing in today's world.

Yet, such a state may in turn motivate the system to achieve a more complex level of organization, by seeking and integrating relevant data of which it had previously been unaware or had deliberately ignored. As we have equally noted in the Preface, it is for this reason that the present global circumstance, with its increasingly acute awareness that things simply cannot go on as before (without a major breakdown in all of our life-support systems) should be seen as an opportunity for, rather than an obstacle to, radical transformation.

For Quantum Relations theorists, just as for general systems thinkers, it is imperative to recognize that the feedback processes are constant features observed throughout the cosmos. One can equally perceive them in organic, suborganic, and supraorganic worlds, from atoms to social groups to the biosphere of our planet. In such fields as anthropology, biology, information and communication, economics, environmental sciences, humanities, neuroscience, psychology, physics, and sociology, open systems of a biological, electronic, social, psychological, and hermeneutic nature can be treated non-reductively in terms of nonlinear concepts such as dynamic wholeness, mutual causality, feedback loops, self-stabilization and differentiation, information flow, and transformation. Consequently, the nonlinear models pioneered by general systems theory and cybernetics are further developed and refined in many of these branches of learning, and QR theory has in turn extensively and profitably employed them in its technological applications.

In the life sciences, contemporary developments and refinements include nonlinear concepts such as Ilya Prigogine's "dissipative structures," Humberto Maturana and Francisco Varela's "autopoiesis," Stuart Kauffman's "edge of chaos," Per Bak's "self-organized criticality," Niles Eldredge and Stephen Jay Gould's "punctuated equilibrium," James Lovelock and Lynn Margulis' "Gaia hypothesis," Benoit Mandelbrot's "fractal geometry" and, generally, the "mathematics of complexity" that is associated with process ontology, dynamic systems theory and is based on nonlinear equations. Together they form a body of scientific thought called chaos and complexity theory (or "chaoplexity") to which Quantum Relations equally belongs and which may lead, in Fritjof Capra's words, to a "unified theory of living systems" (Capra 1997, p. 154f).

The Gaia hypothesis is particularly relevant to Quantum Relations, because it looks at our entire planet as a very complex, self-organizing system, which could, in fact, be successfully modeled and enhanced with QR-based computational tools. James Lovelock first proposed this hypothesis in 1979, and then developed it throughout the 1980s, together with Lynn Margulis. In his words, the Gaia hypothesis should be considered "as an alternative to the conventional wisdom that sees the Earth as a dead planet made of inanimate rocks, ocean, and atmosphere, and merely inhabited by life. Consider it as a real system, comprising all of life and all of its environment tightly coupled so as to form a self-regulating entity" (Lovelock 1988, p. 12).

The idea of self-regulating systems comes from general systems theory/cybernetics, but the Gaia hypothesis uses it innovatively, by linking together, through a large number of feedback loops, both living and nonliving systems such as plants, animals, humans, rocks, oceans, and atmosphere. For Lovelock and Margulis, Earth is an immensely complex, self-organizing system in which life generates the conditions of its own possibility against almost impossible odds (at least in terms of the second law of thermodynamics). In turn, evolution cannot be limited to adaptation or to a static environment. As Lovelock puts it, "So closely coupled is the evolution of living organisms with the evolution of their environment that together they constitute a single evolutionary process." (Lovelock 1988, 99)

The Gaia hypothesis shifts the focus of scientific research not only from evolution to coevolution, but also from adaptation, survival of the fittest, and random variation to mutual dependence, creativity, and cooperation. In other words, it assigns a central evolutionary role to symbiotic processes. Capra, commenting on the theory of symbiogenesis–a hypothesis first elaborated by Konstantin S. Merezhkovsky and then, over half a century later, by Margulis – points out that all "larger organisms, including ourselves, are living testimonies to the fact that destructive practices do not work in the long run. In the end, the aggressors always destroy themselves, making way for others who know how to cooperate and get along. Life is much less a competitive struggle for survival than a triumph of cooperation and creativity. Indeed, since the creation of the first nucleated cells, evolution has proceeded through ever more intricate arrangements of cooperation and co-evolution" (Capra 1997, p. 238).

Starting from the Gaia hypothesis, Capra proposes the metaphor of the "web of life" for describing Earth as a vast network of interacting and mutually supportive self-organizing systems. This metaphor is grounded in mutual dependence and cooperation at all biological and cultural levels, rather than in a raw "struggle for life" and "survival of the fittest." As Margulis and Sagan put it, "Life did not take over the globe by combat, but by networking" (Margulis and Sagan 1986, 17). Moreover, in the general systems view that Capra shares with the proponents of the Gaia hypothesis, life is far from limiting itself to survival and reproduction. Rather, it exhibits an "inherent tendency to create novelty, which may or may not be accompanied by adaptation to changing environmental conditions" (Capra 1997, 221).

Within a planetary framework, it should be clear that a nonlinear evolutionary model such as the "web of life," which QR theory equally subscribes to, is more appropriate for learning and research than the reductionist models of Western mainstream life sciences. The nonlinear evolutionary models encourage and support a cooperative, symbiotic view of evolution, in which all living and nonliving components of the overall global system depend on each other for their wellbeing and sustainable development. They also stress creativity and diversity, rather than uniformity, as key factors in both natural and cultural evolution. Most important of all, they stress the great responsibility that each self-aware member of the global system has toward other such members, as well as toward life in general.

Finally, the web of life and symbiogenesis are ideas that certainly do not belong exclusively to Western culture. They can equally be found, in one form or another, in all the major cultural traditions of the world. Consequently, it would constitute an excellent point of departure for global, intercultural research and dialogue. *Global Intelligence and Human Development* (Spariosu 2005) discusses at length the theoretical advantages of general systems theory and its offshoots, the theories of complexity and self-organization, over their scientific, reductionist counterparts, especially within a global reference frame. That book also points out the close similarity between the nonlinear views of these theories and those of early Buddhism and Daoism. Here we can only sum up, very briefly, the basic views of the last two systems of thought in relation to the Quantum Relations Principle.

Reciprocal or mutual causality is a cornerstone of Gautama Buddha's thought and practice, as revealed in his teaching of "dependent co-arising" or "dependent origination" (*pattica samuppada*). The Buddha changes the traditional, essentialist Vedic definition of causality to express dynamic relationships rather than substance. For instance, in Part II of the *Samyutta Nikaya*, called Nidanavagga or the Book of Causation, the Buddha explores the *nidanas* or "causes" of suffering, such as ignorance, volitional formations, consciousness, name-and-form, and so on, by asking his auditors: "And what, bhikkhus [monks], is dependent origination? With ignorance as condition, volitional formations [come to be]; with volitional formations as condition, consciousness [comes to be]; with consciousness as condition, name-and-form; with name-and-form as condition, the six sense bases; with the six sense-bases as condition, contact; with contact as condition, feeling; with feeling as condition, craving; with craving as condition, clinging; with clinging as condition, existence; with existence as condition, birth; with birth as condition, aging-and-death, sorrow, lamentation, pain, displeasure, and despair come to be. Such is the orgin of this whole mass of suffering. This, bhikkhus, is called dependent origination." (Samyutta Nikaya, II. 1)

The Buddha is not interested in finding out what "causes" produce a given factor *A* (such as suffering, ignorance, craving, clinging, and so forth). Rather, he seeks to determine what else happens in relation to the happening of *A*. In this sense, the occurrence of *A*, say, craving, provides a locus or context in which *B*, ignorance, can equally occur. Or, put in another way, *B* arises co-dependently with *A*. So, ignorance is not the "prime cause" of suffering, as linear interpretations of Buddhism would have it. Rather, it is present with other factors when suffering occurs.

Nor should the relation of *A* and *B* simply be understood as a contiguity of events, as in David Hume's view, which Western scholars have occasionally mistaken for the Buddhist notion of causality (Macy 1991, 53). For Hume, physical events flow past us, but are objectively unrelated, and it is only our mental operations that infer a causal connection between them. By contrast, the Buddha perceives both an ontological and an epistemological connection between events, so that mutual causality is the way things happen, or the "nature of things" (*dhammata*). As we have seen in previous chapters, Quantum Relations Theory is based on the same principle of mutual causality.

The Buddhist view of subjectivity or human agency applies to both general systems theory and Quantum Relations: it dissolves the subject/object dichotomy into a reciprocal play of relations between fluid variables. The linear causal paradigm in both Vedic India and the modern West exclusively emphasizes either the perceiver or the perceived in the cognitive process. The Western empiricists, for example, believe that the world is the cause of our perceptions, registering its data on neutral sense organs. Philosophical solipsism, on the other hand, holds the symmetrical opposite view that external phenomena are our own projections. By contrast, the Buddha denies neither the being-there of the sense objects nor the projective tendencies of the mind, but regards the two as mutually conditioned: we both shape and are shaped by sensory experience – an insight that QR equally shares, along with general systems theory.

The concept of dependent origination or mutual causality is equally present in the three major strands of traditional Chinese thought, namely Daoism, Confucianism and Buddhism (the last one came to China from India, during the first century A. D.). Daoism is believed to be the most ancient, its founding being attributed to the poet-sage Lao-Tzu (604–531 B. C.), an older contemporary of Confucius and Gautama Buddha. Although early Buddhism and Taoism obviously developed separately and in parallel for over five hundred years, they have many points in common, including the notion of the cosmos as process, or as a web of dynamic relations (rather than independent substances or identities).

Lao Tzu's nonlinear way of thinking, moreover, just like that of the Buddha, will also undergo essentialist interpretations later on in the tradition, at the hands of both Eastern and Western scholars. But, most important from a contemporary global perspective, both ways of thinking share a nonviolent, peaceful approach to the human and natural worlds, as well as to the web of life in general, which we have equally adopted in Quantum Relations theory. For the purposes of the present argument, we shall very briefly look at early Daoism, although both Confucianism and Chinese Buddhism would yield important insights regarding the kind of thinking which works well in the current global circumstance and which we have also adopted in our consulting work based on the QR principle.

The Chinese word "dao" is roughly equivalent to the Sanskrit "dharma" and can, like the latter, be translated into English as "path," or "way," or "nature" (of things). Early Daoism, just like early Buddhism, is a practice, more than a philosophy, being concerned with the most appropriate ways of living, based on a proper understanding of the nature of reality. This reality is seen as process or as continuous flow, where all things arise and fade through the formation and dissolution of interactive networks of relationships, based on reciprocal or cyclical causality. Like early Buddhism, Daoism proposes a way of emancipation for the individual, not so much by escaping the cycle of mutual causality as by realizing its nature and becoming one with it. The enlightened Daoist practitioner, just as his Buddhist counterpart, is able to ride up and down the billow of existence with steady equanimity.

Nonlinear causality in Daoism is expressed in terms of a ceaseless play of polarities. Just as in early Buddhism (and in QR and general systems theory), polarities in ancient Chinese thinking are metaphorical and paradoxical ways in which one can describe fluid relationships among elements with ever-changing identities. Ancient Chinese thinking, therefore, no less than its early Buddhist counterpart, "entails an ontology of events, not one of substances" (Hall and Ames 1987, 15). In this kind of process ontology, human events and agencies are not understood in terms of qualities, attributes, and substances, but in terms of specific contexts and relations. As such, process ontology "precludes the consideration of either agency or act in isolation from the other. The agent is as much a consequence of his act as the cause" (Hall and Ames 1987, 15). No less than early Buddhism, then, early Daoism makes no substantialist distinction between subject and object, but operates instead with a dynamic process of fluid interrelationships, based on mutual causality.

Although early Daoism plays with polarities, it is definitely not polar or binary-oppositional thinking, based on a mentality of power, as some Western and Eastern scholars seem to believe. In Quantum Relations vocabulary, the Daoist emancipates the "weaker" pole from the feedback cycle in which a power dialectics has locked it and where, in due course, "strong" inevitably becomes "weak," and "weak" becomes "strong." Then, the practitioner allows each of these poles to create its separate amplifying feedback loops. In these new loops, the strong or the violent will not become weak, but will simply breed more of the same:

> One who assists the ruler of men by means of the way does not intimidate the empire by a show of arms.
> This is something which is liable to rebound.
> Where troops have encamped
> There will brambles grow;
> In the wake of a mighty army
> Bad harvests follow without fail. (Lao-Tzu, I. XXX)

War will bring with it more war, sowing destruction rather than bountiful harvests for friends and foes alike. If one must use arms, one should do so without relish and without gloating over a defeated enemy. Victory is as undesirable as defeat, because both are part of the same destructive feedback loop:

> Arms are instruments of ill omen, not the instruments of the gentlemen. When one is compelled to use them, it is best to do so without relish. There is no glory in victory, and to glorify it despite this is to exult in the killing of men. . . . When great numbers of people are killed, one should weep over them in sorrow. When victorious in war, one should observe the rites of mourning. (Lao-Tzu, I. XXXI)

The Daoist practitioner does not try to change the world by force, for he knows that force will result in more force. Therefore, he "avoids excess, extravagance and arrogance" (Lao-Tzu, I.XXIX). At the same time, he cultivates meekness, which will in turn generate more meekness. One should treat all people with kindness, indifferent of their nature. In this way, one will amplify the feedback loop of kindness, instead of engaging in a conflictive mimetic relationship and thus relapsing into a destructive feedback cycle:

> Those who are good I treat as good. Those who are not good I also treat as good. In so doing I gain in goodness. Those who are of good faith I have faith in. Those who are lacking in good faith I also have faith in. In so doing I gain in good faith. (Lao-Tzu, II. XLIX)

This short detour through the precepts of the ancient tradition of wisdom or the *philosophia perennis* as Leibnitz dubbed it, shows its great relevance to our own age, when the planetary framework of globalization implies different rules and principles of human interaction that need equally to be remapped and reorganized at the local and regional levels. To give just one example, if we allow Buddhist and Daoist thinking to resonate with us, we can revisit the notions of feedback loops, as systems theorists and cyberneticists currently conceptualize them. Thus, some of our QR computational models that feature human relations that are appropriate within a planetary framework have replaced the technical terms of "positive" and "negative" feedback loops (which in general systems theory seem somewhat confusing and counterintuitive) with "constructive" or "life-enhancing" and "destructive" feedback loops.

Through these terms we have, like the Daoists, decoupled the asymmetrical polar structure of many of the feedback circuits. Consequently, only the dominant pole remains within the reference frame of power, with its ceaseless interplay of opposing forces, whereas the "weaker" pole becomes the organizing principle of a different mentality altogether. In this alternative frame of reference, the notion of equilibrium (of opposing forces), no less than that of a "state far from equilibrium," can be rendered inoperative, so that different, nonviolent concepts of and relationships among self-organizing systems and other phenomena could emerge over time, at all levels.

We have redefined resonance accordingly, as an amplifying type of feedback loop, instead of a self-regulating type. This has allowed us to understand how self-organizing systems interact with other such systems, whether they are nested within or host those systems; how certain "local" characteristics can spread so rapidly throughout systemic wholes; and how apparently insignificant events may trigger catastrophes (in René Thom's sense) within a given system or polysystem.

In our QR models, we have also contrasted the notion of resonance with that of mimesis in relation to constructive and destructive types of amplifying feedback loops. Mimesis, translated as "mimicry" or "imitation," often involves an asymmetrical relation of power that can lead to amplifying destructive loops. It can therefore be seen as mostly a destructive form of amplifying resonance that feeds on conflict. In turn, there are cooperative or symbiotic forms of resonance that generally lead to productive types of amplifying loops, at least in reference frames that are not power-oriented. All of these concepts, adapted from Buddhism and Daoism, have helped us understand and model how a planetary world based on nonviolent and peaceful systems of values and beliefs could emerge in the near future.

To sum up our arguments in this chapter, our QR-models are based on the "web of life" in its most diverse and complex aspects, including human relations and interactions. Unlike reductionist scientific theories, which generate reductionist technological platforms, QR implicitly acknowledges diversity and alterity as the very conditions of existence. Whereas the reductionist theoretical models perpetuate the hegemonic pretensions of mainstream Western science, attempting to impose its dualistic, Cartesian perspective on all cultures in the guise of objective, universal knowledge, DFO/FOR models can take into account and process widely different cognitive perspectives, including linguistic, philosophical, cultural, sexual, and other observer-dependent variables. At the same time, they can continually and automatically update, reframe and reorganize their data as new global realities emerge.

Furthermore, like other contemporary strands of general systems theory, QR acknowledges that hierarchies as modes of organization are best understood not as "centers of command and control," but as reference frames or levels of complexity embedded or nested within each other and engaged in constant communication and mutual interaction. QR theory thus supports and enhances a cooperative, symbiotic view of our world, in which all living and nonliving components of the planetary system and subsystems depend on each other for their wellbeing and in which each perspective needs to be acknowledged and respected as potentially valuable for the common good.

In conclusion, QR-based technological platforms are built in line with what we would like to call the emergent ethics of global intelligence and planetary wisdom. In our view, this ethics ought to be based on a mentality of peace, defined not in opposition to war, but as an alternative mode of being, feeling and acting in the world, with its own system of values and beliefs, and reference frames. We call it "emergent" because it depends, no less than global intelligence and wisdom, on continuous intercultural dialogue, research and cooperation. It should also imply a second-order ethical system. Unlike a first-order of ethics, in which one accords preferential treatment to one's own group, be it family, tribe, nation, religion, etc., a second-order of ethics is based on the Golden Rule (treat all human and other beings as you would like to be treated yourself) and is, therefore, the most appropriate mode of thought and behavior to adopt within a planetary reference frame.

By the same token, the quality and nature of the QR applications in a local, regional and global framework will obviously depend on the quality and nature of the intercultural databases that they will draw on and, above all, on the mentality and principles that will inform the collection and processing of such databases. In this regard, we ought to insert a note of caution that pertains to the development of any new technology. Although QR theory is implicitly and naturally sympathetic to a symbiotic, cooperative approach to the web of life and could lead to the creation of exciting technological applications in that spirit, it could also lead, like any other theory, to less desirable applications that will have unintended, distressful consequences for future human development.

The DFO/FOR model provides for large-scale computation of observer-dependent properties, whether they belong to observing and observed systems that are configured within power-driven frames of reference or those of any other organizing principle. For example, QR-based technological platforms can also easily implement some of the NSA projects in cognitive computing (such as LifeLog) that have now been publicly disclosed in the United States and the international community. Although such EI technological platforms are certainly useful in tracking and monitoring terrorist and other criminal activities, they can also be improperly used by an overzealous (or politically motivated), information-gathering, governmental agency, or even by a criminal organization involved, say, in cyber terrorism.

Consequently, it is a particular human mentality that will determine the uses of QR technology, and not the technology itself that determines this mentality and the type of life-enhancing or self-destructive feedback loops it will generate. It bears stressing, again and again, that whether we move from the digital age to the quantum age and beyond, the future of human development will depend less on advanced technological platforms and much more on our collective ability and, above all, willingness to shift to a mentality and practices that are oriented toward global intelligence and planetary wisdom. As we have emphasized in the Preface of the present book, this is an arduous task, involving much preparatory work on the part of all of our world communities, consisting primarily, and most importantly, in developing local-global, intensive learning environments, supported by advanced technological platforms that are based, in turn, on QR or similar principles. It is for this purpose that in Part Two, Chapter 8 below, we propose the electronic architecture of a planetary network of intercultural centers for integrated knowledge, technology and human development that would support such intensive global learning and research environments.

Part Two.
The Quantum Relations Technology: Global Applications

Chapter 5.
The Operating System of the Quantum Relations Technology: The Interactive Search, Analysis and Transaction Environment

To index and deliver content and information is a simple and mechanical task. But, to locally and globally identify, analyze and deliver the right content to the right user at the right time is a very hard problem, indeed. One of the most exciting features of our Quantum Relations Technology (QRT) is its *interactive, automatic system of search, analysis and delivery of content* that is fully capable of the kind of advanced functionality of matchmaking and evaluation that is expected of future intelligent technology. This system renders obsolete the current Google and Yahoo/Microsoft search and transaction technologies and can revolutionize the whole field of the management of intelligent, interactive, real-time information and content.

In this chapter, we shall address only tangentially the standard server operations with their general transaction capabilities, because those are rather well understood and basic in their implementations, which simply require the best common practice. Instead, we shall focus on the QRT-native, search and analytics engines that enable the central DFO/FOR paradigm in its core. These technologies actually constitute the foundation for all the applications that are implemented on the QRT core system (see Chapters 6, 7 and 8 below).

Background of the Search, Analysis and Matchmaking Functionality

Today's information and transaction engines and the intelligence tools that are used, for example, in e-commerce solutions, have so far failed to evolve towards each other. With a few exceptions, they occupy completely different universes. Increasing demand for more complex (and more narrow) search-and-delivery applications has led to significant developments in the algorithms used by search-engine providers such as Google and Yahoo (now Microsoft). Nevertheless, the functionality of such search engines inside online commerce, for example, is only marginally satisfying in personal use and even less so in business-to-business (B2B) and in real-time, push-content environments. Business intelligence tools, on the other hand, provide somewhat better results in terms of searching for commercial information, but their approach is out-of-date, inflexible in application over a wide range of problems, and nearly never dynamic or genuinely real-time.

What is in fact needed is, on the one hand, a dynamic, real-time approach to the search and delivery of information and content, and, on the other hand, retaining the information learned in one search, transaction and delivery so that it can be used in subsequent or even parallel transactions. Furthermore, the transition from information to usable knowledge must be managed in the background and automatically. This implies local search and delivery tracking that results in high-end, real-time knowledge applications that consciously adapt to users' specific information-space and knowledge targets.

We want to be real-time and we want to filter out information that is not relevant to the particular users of the search and delivery process. Such new and advanced platforms need to include pull-and-push technologies as a basic strategy to provide fully interactive search flows, taking advantage of Enhanced Intelligence (EI) tools. In other words, we need an intelligent *foundation operating system* that is not only the foundation of search, analysis and match-making, but also that of any computational technology of the future. Every new transactional tool must become an intelligent extension of the same core foundation standards.

The next generation of search-and-delivery applications must 'know' their users, understand what they need, and must treat information gathering and analysis as a continuously evolving process that 'shadows' the user's specific behavior. Perhaps most revolutionarily, they must also anticipate the user's future knowledge and needs. In fact, they must anticipate new and evolving users.

Redefining the term "user"

Search engines must be able to understand and serve users as human individuals, groups of humans (self-defined, or system-defined by profile or similarities of detected behavior), machines, groups of machines, bots, or individually detected, instructional codes in networks and software. The term "user" in this context is anything that can interact with the QRT system. A user can also range from a single individual person up to the entire global society, together with all its technology and sub-behaviors. Any similarity of behavior or of collective needs is potentially a "system-defined or detected user group" that is automatically shadowed by a profile, becomes continuously analyzed and can be adequately served when it begins to interact with the QRT system.

The Weakness of Current Systems

A simple example will illustrate the weakness of current search systems: Suppose I am searching the term "Cobra," using Google. All of the top Google hits relate to the U.S. Consolidated Omnibus Budget Reconciliation Act (COBRA). The next set of hits relate to Cobra Electronics Corporation, Cobra Golf, and Cobra Motorcycle Exhausts. It is possible to narrow the scope of my search, but why cannot my search engine be smart enough to know that I don't care about U.S. insurance and am certainly not in the market for motorcycle exhausts? Why should my search engine be indifferent to my needs and too unsophisticated to learn my likely interests, despite the fact that I perform dozens or hundreds of searches every day?

Q-Search: The QRT Standard Search and Analysis Component

A significant paradigmatic shift in the definition and development of technologies of search, analysis and delivery of content is therefore needed to transform current search engines into real-time, dynamic and interactive systems. Our Q-Search engine represents just this kind of paradigmatic shift: it can establish, through dynamic models of quantum relational context, a unique, n-dimensional, "universe" for each authenticated user, user groups, or adaptive and recallable user profiles that fit the probabilities of a specific request.

The Basic Features and Benefits of the Q-Search Engine

The Q-Search technological platform delivers all the functionality of today's "state of the art" interactive content technologies. In addition, Q-search mechanisms and queries are flexible and user-definable, both qualitatively and quantitatively, performing such functions as:

1. Organize, identify and make permanently traceable searchable content through global IPv6 technology.
2. Not only "remember" all content that ever existed and the Q-Search system was exposed to, but also keep a copy of such content, track changes of it and store them as knowledge memorized in time-series values and /or flat files for future reference and change analysis, so that the dynamic of content can be used intelligently in the search process, directly or in context.
3. Not restrict search content only to documents, but also include packages of structured and unstructured data such as audio, video, graphics and pictures, text, tables, numbers, various media, excerpts of content in any

combination and composition, and even other user or system-defined packages. Again, it can do so, by preserving histories of contextual hierarchies and n-dimensional affiliations.

4. Package all user-defined content packages with an Ipv6 identifier with user-defined persistence.

5. Develop search interfaces so that they become fully integratable and embeddable objects into any IPv6 web-enabled device or technology, including hidden devices and submerged data pipelines, as well as any other content-delivery, interactive and reactive, push-and-pull device.

6. Feature a fully interactive interface to classical IPv4 browsers as plug-in extensions for Microsoft Explorer, Firefox, Safari, and other popular architectures.

7. Execute content-searches that are continuously redefined, changeable and recallable through search spaces that are user or profile/context invoked and defined; the system is designed to continuously learn from the user and others like him/her/it (or any other definition of "user" as given above), proactively updating the user's search-space in anticipation of specific needs.

8. Permit, through its search interface and backend technology platform, abstract search goals that service complex, user-defined spaces such as "sentiment", "opportunity" and "risk".

9. Enable from the beginning full commercial and other real-time applications such as "Business Intelligence" and "Continuous Situation Awareness."

10. Suppress back-tracing to the user by a system operator or advertiser on the Q-Search system, because all profile content is embedded into an encrypted avatar agent and not locally; Q-search, therefore, enhances user privacy by connecting the actual user with his "System Avatar" through bio-encryption technology; this makes Q-Search a very secure environment with concurrent enablement of true personalization and dynamic optimization of the user profile inside the system.

11. Allows the Q-Search system to "understand" and continuously update the user's level and type of skills, abilities and comprehension, in order to select the most appropriate and understandable content, while at the same time enhancing the search by offering the next level of skill or comprehension. The search technology can therefore help continuously and naturally expand the user's knowledge and skills, without additional effort or pressure to do so.

Thus, Q-Search is the first technological platform that offers to transform "searching the Web for content" into valuable pull-and-push content applications and embedded technological components for any content-carrying, content-receiving or content-broadcasting devices, ranging from casual information and entertainment to real-time business intelligence and from global, real-time, commercial support to global end-to-end applications. Indeed, Q-Search is able to drive the entire global information-and-content backbone of the online environment, if applied on that scale.

The Problems To Solve

To understand why Q-Search represents the best strategy of content management for the near future, one should understand the severe limitations of the existing browsers and the search-and-delivery components of their engines. These limitations include:

1. they do not utilize the potential of IPv6
2. they are static
3. they are strictly text-search based
4. the search 'engine' and browser designs are not well integrated
5. they are not total immersion experiences
6. they are not well customized to individual users and real-time situations
7. they cannot search for abstract ideas or risk/opportunity spaces
8. they cannot deliver active and real-time business and other intelligence
9. they are not built to support demanding, dynamic business and other environments

Furthermore, current search engine technology is limited to discovering documents based solely on literal searches. They may allow the possibility of the query being expressed in something approximating English or some other natural language, and they may have some understanding of synonyms or, possibly, even of topics; but, fundamentally, all they do is return objects whose content contains only what was literally requested.

As an example, take any search engine and try searching for something like "today's good news." One might wish to receive results that include news stories reported today and scored as positive by some rating authority (which may be the user's own set of preferences as encoded in some functional avatar). However, the results will instead include documents that contain the words "today's," "good," and "news". Indeed, most of the top results refer to religious or spiritual sites whose page title is literally "Today's Good News," but have nothing to do with positive news stories being reported today.

So, while the current crop of search engines might excel at some tasks, they fall woefully short at others. Moreover, the current direction of development in search technology seems to be limited to discovering more of the Web to index, to improvements in the query language such as better natural language support, and to improvements in the page ranking algorithms, which can at best reorder or augment the existing result sets. In other words, there is nothing that the current generation of search engines can do or is even attempting to do that addresses anything other than literal searches.

In sum, we see four main problems with the search technology of today:

1. Searches are based on literal content. Typically, only words and phrases are indexed, and only in rare cases a general topic classification might also be included.

2. The frame of reference of the user is completely ignored. Perhaps a search engine ranks recent documents as being more relevant, in which case there is a vague notion of the present time. But imagine more interesting user context, such as political affiliations or gender. Let us assume that I am a politically conscious woman who lives in the USA and might want to know which candidates would best represent me in the next election: it would be foolhardy to expect a reasonable answer from a current search engine. It doesn't really help even if I provide context, such as: "I'm a liberal woman who votes Republican; who are the candidates in 2016 favoring abortion and gun control?" Indeed, at least one search engine returns a link about a progressive Democratic candidate first.

3. Perhaps even more importantly, there is no dialog with the system, no chance to refine the results toward a user's content interest, the only exception being a not always reliable – and often frustrating – job of correcting misspellings or grammar in the English language.

4. Content is delivered to a user only at the particular moment that the user initiates the search activity. But, the user could have searched for that content last week, when it was not available yet. As the content emerges on the global networks, it should be at least brought to the user's attention through embedded push technologies, should the user wish to receive dynamic updates on his general interest topics and specific requests. Nevertheless, there is not one provider of such technology today, nor has anyone envisaged such capabilities on a global content-delivery scale (except for our Q-Search which uses such delivery methods inside its Q-Avatar Technology®).

Now let us consider the basic architecture of web search engines to see why these shortcomings might be there. Very broadly, there are three components to any web search engine: a web crawler, a content indexer, and a query engine. The web crawler is responsible for fetching content from the Internet. The indexer maps words and phrases in the content back to their source. This map, known as an index, often might include weight data, such as number of occurrences and relevance in each document, whereas the query engine handles the language processing of the user's request and does the lookup, using the index.

One could argue that today's web crawlers do a reasonable job of discovering content. Major search engines claim upwards of 20 billion indexed "web objects". But the majority of these focus on static content – as if the Web were a giant unorganized library, and there were fierce competition among search engines to claim who knows more about the books in this library. Yet, in many essential ways, this library model of the Web is not the correct one, because it constantly suffers from being out of date. There is often no distinction between older and newer content, so when doing a search, say, for a product comparison, you are much more likely to get the one with the most links to it, even if it concerns discontinued or obsolete versions of the product.

In recent years, the content-gathering phase of search engines has expanded to include not only crawling the web and "pulling" information, but also some data feeds that "push" information to the search engine, such as news. Unfortunately, this time-sensitive data seems to be treated more or less like any other data in the "library;" or even worse, it is sometimes arbitrarily given a higher-relevance score for the first few hours or days after it has arrived.

Let us now look at the query engine. Here again, there is competition among current search engines to provide natural language input and other clever features, such as the Google Calculator, for example. But these are still mostly limited to a single input, producing a single result set. The exceptions are in the realm of offering a suggestion with a spelling correction or to search for documents in some other domain of the search engine, such as books or news. Refining a search is almost always nothing more than adding additional "terms" to the existing query – which is still the one-query/one-response paradigm.

Finally, let us return to the indexing process. Despite appearances, this is perhaps the most severely limiting component of current search engines. The content indexer takes each document and treats it mostly like a bag of words. It then maps each word in the bag back to the source of the document. Nearly always, this mapping contains additional information such as the number of occurrences and also distances from other words in the same document. Furthermore, there is information about where in the document the word was found, not only positional, but also structural. All of this additional information is used to help rank the relevance of the word in the document and is one of the inputs to rank the overall document in the result set. Let it be said that the results are often impressive. Then, how is this limiting?

Well, no matter how much context is associated with each word in the index, in the end only words are indexed, and this limits one to literal searches. We acknowledge that synonym search is also available, but this is only an incremental improvement at best. And, obviously much worse, a major search engine actually shows a top hit of "negative" when searching for synonyms of "positive". So if I would like documents that are positive in tone, a literal or

synonym search cannot possibly help. Consequently, lots of additional linguistic analysis can and must be done at the time of indexing to allow much
richer result sets that address what users mean to search for, especially when just
the right word cannot be found.

The Ontological Solutions

Q-Search is a problem-solving technological platform that operates broadly in
three domains relevant to search issues. It excels at real-time content acquisition,
sophisticated real-time data processing with an emphasis on linguistic content
analysis, and on-demand query, computation, and delivery. In addition, Q-
Search offers advanced monitoring and management tools to keep data-centers
aware and provide flexibility in information, product and content offerings.

For the past several years, our Quantum Relations Technology (QRT) has
been collecting streaming data, structured and unstructured, from a wide-array
of sources, including high-quality, major news wires. All of this data is captured
and recorded at millisecond resolution. The system is also able to automatically
fill gaps that can occur due to network or provider outages. QRT also monitors
lower-frequency feeds using RSS and ATOM. These strengths overcome the
limitations of web crawling and keep the content database fresh and relevant.
The advanced linguistic processing of data on a real-time basis is perhaps the
core of QRT. One can think of this processing as a type of indexing, but it goes far
beyond mapping words and phrases to documents. Each document or content
object is analyzed across hundreds of psychological, linguistic and subject-
related vectors.

Advanced statistical treatments are also done on each document, ranging
from Bayes to Latent Semantic Analysis to advanced neural networks such as
Restricted Boltzmann Machines and proprietary genetic searches. All of this is
done in real-time as hundreds of thousands of new documents and content
objects arrive each minute. QRT can classify each dataset (documents, fragments
of documents, pictures, videos, media, etc.) in virtual vector spaces with billions
of possible dimensions, each dimension being native to specific frames of reference under which such data set could be looked at or searched for.

An important point to understand regarding linguistic processing is that its
goal is not to determine what a document means in any absolute sense. Indeed,
the Quantum Relations Principle, on which the existing software is based,
postulates that there is no absolute meaning, but a different meaning in each
Frame of Reference (FOR). Since meaning is based on one's FOR, QR engines
extract information from documents that only becomes meaningful within the

context of some frame of reference and where multiple meanings emerge naturally.

Given that documents now live in a space where context is not just important, but required, the Q-Search query and delivery technologies are able to provide results that go far beyond what is possible in the current world of literal search. Furthermore, results no longer represent a point-in-time view of the world, which is mostly static. For example, queries like "show me the price of a Zune on eBay" can be returned in a time-series format that shows how the prices have evolved over time, and where new results automatically update. In fact, if there is a detectable pattern in price development (e. g., season-related or other price-influencing factor), the query may even return expected price behavior in the future and help the content searcher to make a better timing decision on when to commit what resources to what content or product. This transforms search from returning data or information into a source of real-time knowledge or actionable commercial intelligence.

Using Avatars in Smart Content Mapping

We have pointed out that one of the main shortcomings of current search engines and content-tracking technologies is that they are not familiar with the search preferences and the search-and-content history of their users. One way of thinking about this is to imagine the system as having a large number of "preferences" which can be set by the user at runtime. This way of thinking is cumbersome and leads to impossibly complex and hard-to-configure software.

A better way of thinking about it is to imagine the search space of the Internet as a "universe" in which we have an avatar to whom we can give our query. The avatar is, in a very real sense, a local copy of ourselves who knows much about our needs and our history, and can use that information to narrow the search results to information that we would find useful.

This is similar to, but also distinct from, the concept of agent-driven software. We are furthermore proposing here that the avatars of different people can interact with each other, if both users explicitly give such permission (thus taking the context of social networking software such as Facebook to the next level). There is no reason, in principle, why they could or should not give it, considering the extensive security features embedded into Q-Search. In fact, our avatar is an intelligent software agent uniquely skilled at traversing the searchable universe and returning with the data that we have requested. Nevertheless, such avatar could, if given permission on both sides, contact other avatars for knowledge or collaboration, should a "search profile" favor such interaction (this could also mean, for example in e-commerce, that several avatars can "negotiate" for a better price with a content provider, when acting as a group and negotiating in volume).

To take the Cobra example above, our avatar, when asked to find "Cobra," will realize (knowing our history) that we are likely to be searching for information about classic, high-performance Fords, and not about motorcycle exhausts, U.S. insurance law, or even venomous snakes. Likewise, when asked about today's good news, our avatar will return stories about rescue workers saving kittens, and not about Christian devotionals (or, for other users, the reverse, as the case may be). This is achievable because we design each of the phases of the search engine technology (web-crawling, indexing, and query processing) to build and manipulate objects in the searchable, dynamic universe that also represents the user's contextual environment. The mechanism that holds all of this together, the "glue" of the searchable universe, is Quantum Relations Technology, which essentially provides a dynamic, flexible, and comprehensive way of organizing and manipulating data of any kind.

As Microsoft itself pointed out, "more than 25 percent of all search queries aren't about discovering new information at all – they're meant to navigate to information and websites that people have already visited." (Technology Review, at http://www.technologyreview.com/web/32281/?p1=A2&a=f [August 8, 2011]). Microsoft admits that it does not have a good solution to this problem, because search engines do not store histories, although they have "some idea" of them.

In 1997, the ill-fated Alta Vista dominated the search engine world, but by 2003 – a mere five years later – Google's speed, cleaner web (and software) design, and higher-quality results had driven Alta Vista so far down that it was no longer even profitable; for that reason, it was sold to Overture and Gator, notorious for their adware and spyware. A subsequent purchase by Yahoo! did not salvage the Alta Vista technology. Thus, Alta Vista is dead and Google rules the search engine world, with imitators Bing and Yahoo in tow. Ask (Teoma software) and Gigablast offer essentially the same services in essentially the same way that Google does, but with less breadth, power, or speed (although Gigablast, unlike Google, at least provides meta-tagging indexing and a few other novel, but not generally useful, features).

Computer technology becomes exponentially more powerful every year. When we began development of our software for Big Data news analysis in 2001, the most powerful computers that we could purchase cost hundreds of thousands of euros, had a handful of processors, and a rather limited memory. Today we can buy more processing power on a €200 graphics card. And yet, search engine technology used by Google and others is stuck on 20th century models. Unlike the Google model, our data universe, the QR universe of searchable and computing content objects, built by our web-harvesting software, is designed for today's supercomputers, using today's networking technologies, for today's demanding users.

We already know that the QR methods work for very large data sets and we know that they scale, because of our pioneering work in automated linguistic processing of full text news and similar projects. Software we have written can index a full text news story in under a millisecond, noting hundreds of features such as location of story events, entity recognition, variation from prior reports of the same event, and so forth, and performing this indexing in parallel for thousands of stories. The indexing process used in news analysis is, in fact, building a universe around each story (a unique and dynamic context) for later analysis, rather than computing final results, as is common in current search engine indexing.

Specific Technical Explanations on Selected QRT Features

This section will briefly describe several technological components of the Q-search engine to illustrate and underpin, for the more technically interested reader, how our solutions overcome the fundamental shortcomings of current content search-and-delivery technologies. Here we assume familiarity with the concept of time series, the concept of variable binding in environments, and the basics of IPv6, as well as the Semantic Web. (For more information on the Semantic Web, please see www.w3.org/2001; http://esw.w3.org/topic/Semantic WebTools; and http://www.oracle.com/technology/tech/semantic_technologies/index.html)

The Static Problem

There exists a class of computable objects, called time-series objects, which can constitute the building blocks of a new paradigm of browser/search-engine technologies. A time-series object is something like a function, that is, it is an object that contains computing methods that are aware of and are capable of change over time. A time-series object as a type of functional operator is evaluated in an environment, and its component variables are either bound within the functional object or can be bound to the environment in which the evaluation is performed (if there are unbound variables within that time-series object).

The simplest time-series object is the null or zero object, namely, a function operating on the environment that does nothing, returning no value and having no side effects; in other words, when evaluated, the null object produces a new environment that is identical to the environment prior to the evaluation. A slightly less trivial time-series object is a constant object: When evaluated in an environment, it returns a value that is another, fixed, time-series object, but it does nothing else. This is slightly more interesting than the null time-series object, because it produces an object and is capable of being used as the operand to (for example) a bind function, which permits setting variables in the environment to values.

More interesting are time series objects that perform some computation. Given, for example, a bound time-series object X, one can produce a time-series object that returns X++, that is, X incremented by one, or it can produce $(X+\tau)$ were τ is the current system time. Such objects can be used, for example, as clocks. Imagine a time-series object S that represents the states of some system with three objects, X, Y, and Z. The three objects (which will not be described in detail here) have properties such as mass, size, shape, and position. One may now have a time series operator Q that "evolves" the system S, perhaps using the laws of physics so that the positions of X, Y, and Z are modified as time evolves from τ to $\tau+1$ in the user's local environment, according to the laws of physics in the user's current environment. In shorthand notation: Q*S[t] -> S[t+1]; where '*' means 'applied to', and the functional application goes from left to right.

In addition to being a data storage object and a computational object, a time series object S_1 can also function as a time series operator, and thus one can have "composed" time series objects S_1 S_2 provided that the rules of composition are unambiguously defined within the local environment. Using this approach throughout the entire content-time-matrix will allow us to model contextual dynamics with content dynamics inside a QR global matrix where every element is dynamically evolving in respect to users and providers of any content or information element inside the matrix.

Using the Advantage of IPv6

IPv6 has many advantages over older network protocols such as IPv4. One great advantage is that it is extremely unlikely, over the next 40 years or so, that one might exhaust network addresses, and regardless of how prolific one is in assigning unique network addresses to network objects and content elements. This is interesting for time-series objects (which are, by their nature, a sort of function: given data objects and an environment, the time-series operators produce a new time-series object that is itself a time series). Each time-series object (including time-series operator objects) can thus be assigned a unique IPv6 address, and the method of computation of a potential browser is to 'compose' these IPv6 objects within an environment that is built of other time-series objects. By using IPv6, all information and content detected and stored at any time and anywhere will have a global network address, so that it will become manageable, in multiple ways and simultaneously, by many agents, engines and users.

Every user, agent, content or information element is also expressed as time series and consists of at least one IPv6 number; it may also consist of many such IPv6 numbers, if the user, agent or content object is more complex and has many identifiable and routable subcomponents.

The ability of IPv6 to perform masked 'broadcasts' gives a method of searching both local and network environments for content and matches it with the right user at the right time. To search for objects that meet some predefined set of properties, one expresses the set of properties as a declarative logic statement that one wishes to have evaluated as 'true'. The expression is itself a time-series object and is thus capable of being evaluated in some set of environments; it is a type of time-series object that, when so evaluated, is capable of returning the value 'true,' provided that suitable bindings are available in the environment in which the object is evaluated. After one composes the 'search expression' as such an object, one broadcasts it to the desired subnet of an IPv6 network, and each of the objects in the matching subnet can evaluate the query. If they evaluate it to be 'true', the originator of the query will automatically be contacted and notified that this object (in this particular environment) satisfies that query.

To use the previous example, the query might be the following: Objects X and Y have collided (i.e., the coordinates of their positions are equal). In other words, assuming a time-series operator $x[S,X]$ that returns the x coordinate of the X object in S, and another time-series operator $y[S,X]$ that returns the y coordinate of X in S, and a third operator $z[S,X]$ that returns the z coordinate of X in S, then the declarative query is (AND $(= x[S,X]\ x[S,Y])\ (= y[S,X]\ y[S,Y])\ (= z(S,\ X)\ z(S,\ Y)))$. The query is broadcast to the relevant IPv6 subnet, and the environments within the subnet that are capable of evaluating the query will do so. In those environments where objects X and Y have collided, the query evaluates to 'true', and the object then responds to the query by signaling this fact. The method of signaling is quite simple: A new time-series object is created which has – for those times where the collision in some collection of environments is occurring – a new object that is the collection of those environments (actually, this would be done as a list of pointers to those systems, with appropriate bindings for time and objects).

In each environment capable of evaluating such queries, the method of evaluation is to perform variable binding, and then, if the statement as received does not evaluate to 'true', to apply an inference engine to the collection of statements, operators, and objects in the local environment. This method parallels and is compatible with the 'Semantic Web' technologies being developed by the World Wide Web Consortium, and the properties of the time-series objects could be expressed in Resource Description Framework (RDF) formats. There would be a one-to-one mapping between RDF property lists for an object and its evaluation format when viewed from outside the evaluation system.

In other words, the time-series operator system can be cleanly embedded within the Semantic Web as envisioned by the World Wide Web Consortium. One can think of the time-series object/operator environments as a type of Web Ontology Language (OWL), although in fact they are more powerful than most such ontology descriptors and, in this embodiment, the inference engines (which are generally thought of as a level above OWL in the Semantic Web context) are coupled with the system.

The system of time-series operators and objects is capable, moreover, of producing side effects, and one side effect could easily be to produce GRDDL descriptors of the local objects, which allows one to publish the object properties in traditional HTML or XML formats.

Moving Beyond Text – The Most Exciting Innovation of Q-Search

As we have already pointed out above, current search engines are nice, but they are text-based and static. By comparison, QR time-series representation (that is, representing data as a sequence of objects over some time base) permits moving beyond simple text. One can think of a movie, for example, as a series of fixed objects (frames) that vary slightly from one to the next, coupled with an interpreter (a time evolution operator) that takes the movie viewer from one frame to the next, as time progresses. Similarly, a sound file is a sequence of noises, produced over time in a fixed order.

Storing this data as computable objects, rather than as fixed images or character strings, has many advantages. The time-series objects are evaluated in environments by simple functional composition. Although this technique is simple, it is extremely effective: The lambda calculus and computer languages such as ML, Lisp, and Scheme use this technique as their basic computation tool. Indeed, Scheme, in its purest form, is little more than functional evaluation performed with a vengeance.

If one has experimented with full-immersion gaming on the Internet, including such games as World of Warcraft or Second Life, one may realize how effective these methods are for presenting lifelike and interesting environments. Each player sees the world from his or her (or its) own perspective, and objects in these games are being manipulated, bought, sold, and traded just like objects in the actual world. The result is intense involvement on the part of the players, far more so than in traditional web-browsing. The 'browser' of the future will be a virtual world in which the user searches, finds, manipulates, and transforms objects, rather than a simple, text-based, data retrieval system. Q-Search technology can easily build the basis for this virtual world for content search, delivery and consumption models.

One other problem with traditional search engines is that they return mostly junk. This is in part because of their technological limitations and in part because the purveyors of junk manipulate the systems to force their material on unwilling users, in a similar manner to the producers of email spam. Unlike e-mail readers, however, browsers are 'one size fits all' and are not tailored to the individual user's requirements. But, from a time-series perspective, this is not a problem. Objects are evaluated in the user's own environment before being presented to him/her/it, and therefore techniques such as Avatar-based, Baysian 'spam-object' filters can remove the junk that is not relevant to the particular user connected to the Avatar. As the user moves from one environment to another, different objects become relevant; they become so, because the new environment evaluates the same functions (i. e., the same time-series operators) differently, with new bindings and with new functional operator primitives. Those objects that are not in the least relevant, say, to writing a paper on the history of the Kings and Queens of England, might become very relevant when searching a movie for an evening's entertainment.

Implementations of Semantic Web Technologies

RDF is rapidly becoming the standard language of the Semantic Web. Companies such as IBM, Nokia, Google, and Hewlett-Packard are adopting RDF open-source frameworks, while Oracle 10 g already supports RDF. Oracle 11 g adds the ability to index and query RDF and OWL frameworks, based on a graph data model. Oracle's graph data model is capable of being expressed, simply and unambiguously, within a time-series functional operator model. In turn, we have already implemented a very advanced prototype language for expressing time-series functional operator objects, in our experimental language TSL (Time Series Lisp). Although not fully functional, the language already incorporates object passing, a lazy expression evaluator (which is critical, because of the potential explosion of possible but never-used objects, produced by any reasonable inference engine), an object compiler, inferential and deductive reasoning engines, variable and evolving environments, and a sophisticated graphical interface. The lazy expression evaluator ensures that only those objects are evaluated whose values are actually required to complete some desired computation.

TSL does not, at this time, incorporate the necessary IPv6 object classes and communication mechanisms, but it does implement a sophisticated 'frame of reference' model that is designed to serve as the basis for both local and distributed computation. TSL was designed as a language within which to express ontologies of the sort that will be required for useful Semantic Web implementations.

The Utility of Time-Series and Semantic-Web Applications

The most famous of the current prototype Semantic Web applications is SAPPHIRE (Situational Awareness and Preparedness for Public Health Incidences and Reasoning Engines) system, developed at the University of Texas. Every 10 minutes, SAPPHIRE receives reports on emergency room cases, descriptions of patient symptoms, clinician's notes, and patient self-reports, integrating these into a coherent view of health care in the Houston area. The system has ontology for several kinds of illness, notably flu-like symptoms, and automatically reports suspicious patterns to the Center for Disease Control and Prevention. Although it was configured for Houston, the system was rapidly modified for use in the aftermath of Hurricane Katrina and succeeded in identifying disease outbreaks of respiratory, gastrointestinal, and conjunctivitis in affected areas much sooner than it would have been possible without it.

SAPPHIRE integrates data from many sources and, according to its supporters, demonstrates an important principle: once Semantic Web systems are configured for a general problem, they can quickly be adapted to many different situations, even ones that were not foreseen when the systems were developed. For SAPPHIRE, the general problem was public healthcare reporting. But, the same technologies can be used, for example, in music and media evaluation, social networking systems (MySpace, YouTube, FriendsOnline, FaceBook, and others).

In turn, the time-series data representation model that we are currently developing gives a formal and powerful way of expressing these types of relationships. The lazy evaluation system (computation on demand only) that forms the basis of the TSL language can make such systems practical and scalable.

Hiding the Mechanical Details

Most users of web browsers do not know or care about HTML. Similarly, most readers of news stories do not know or understand or care about NEWSML, nor does the Center for Disease Control and Prevention need to speak RDF to benefit from the SAPPHIRE system. Browsers and search engines hide all of these mechanical details. In a similar way, the details of time- series computation are hidden from users. A user of our TSL does not have to understand time series, or functional composition, or compiler theory, or even deductive reasoning engines. The user treats the time-series objects as a simple, monolithic object, and the system takes care of all the complex details of producing desired values or pictures. What the time-series model does bring to the computer scientist, however, is a robust formal system, capable of enormously powerful computation and, because of its well-established mathematical basis, capable of being produced without the sort of hideous bugs and contradictions that generally plague ontology description languages.

Our current financial news service applications implement many of the design ideas of TSL, including functional composition of time series and lazy evaluation. These have already given many benefits, including the ability to read and analyze thousands of news stories a minute, on relatively simple server systems, and to distribute the results of the analysis to interested users. When the power of these systems is combined with the power of IPv6 and Web 2.0+ URI objects, the results are revolutionary – a new way of seeing, analyzing, and coordinating data that is not only compatible with existing Semantic Web systems and frameworks, but that goes far beyond them in its ability to perform data mining, multiple system coordination, search engine technologies, and immersive browser experiences.

The Q-Search Engine as The Foundation of All Other QR-Based Technological Applications

The QR-based search engine technology lies at the very foundation of embodying the Quantum Relations Principle inside a machine. It is designed not only to find various content, but also as a methodology of placing all found content into a machine-based, dynamic model of the actual, everyday world. All continuously observed content is placed inside the multidimensional structures of the QR model and can be found in any direct or hidden way. The QR-based search engine requires that we build this abstraction of the actual, everyday world as a fully dynamic and interactive QR data depository that lives in parallel to this world. It will not only allow us to find existing or past things, but also to project how things will be, or what our needs and conditions as users of this system will be in the future, since we, the users are part of the machine-based system, part of its replicated world.

Of course, at no time would we claim that we could create a perfect replica of the actual world, because we could never detect all and every element of its content and dynamic changes. Nevertheless, the more we fill the QR system with information, the more accurate it will become. This fact is in marked contrast to the current, non-QR systems, which cannot cope with the huge flood of data, becoming inefficient under its ever-increasing size, speed and volume of content. QR-applications thrive on data and are, therefore, the future of information technology, because the more content is placed in the system, the faster and more accurate this system will become.

We are already using elements of the Q-Search technology in a number of our current applications. This technology allows us not only to establish the targets of a user's search, but also to track it in n-dimensional and fully dynamic frames of reference or contextual views. In Chapters 6 and 7 below, we shall describe, in some detail, the rationales and the features of two of these QR-based systems: the GVE commercial system and Zoe, our proposed medical system.

Chapter 6.
The Global Value Exchange: Managing Commerce in the 21st Century

Let us begin with a statement that may sound provocative: by 2025 at the latest, there will be no more money to be made in the financial industry, which will thus become obsolete. Why is that the case? Because, by then, all of the current trading instruments will have lost their speculative edge due to total computational transparency, and, without speculation, there is no profit to be made in global finance. Thus, investors will either lose their money or move it elsewhere, or treat investment as a simple custodian service of assets and money. The financial industry might still be called that, but it will have been reduced to a none-speculative custodian of stocks and bonds.

The main reason for the impending demise of global finance, as we know it, can be found in the rapid development of information technology. This statement may again sound paradoxical, because it is the same technology that, at present, gives a huge speculative advantage to those who use it, for example in what is known as "high-frequency trading." Yet, once high frequency trading becomes generalized, economically affordable and available to anyone, no trading agent will have an edge over his competitor and will, therefore, have no gain incentive to offer to his clients. As a study by Macquarie Research (May 2009) shows, quantitative technology can already demonstrate full transparency into the future performance of a financial instrument. Full transparency effectively erases the margin for speculation, as well as the time lag between buy/sell operations and value intelligence. It is simply the undisputed rule of efficiency.

But successful demonstration of market transparency by quantitative technology researchers is only the beginning. The speculative trading of financial instruments will come to an end as technology gains the ability to illuminate the true and appropriate value of goods, services, stocks, commodities, currencies and so on, in real-time. First evidence of this development can be seen as big banks and financial institutions can no longer avail themselves of arbitrage opportunities with the emergence of real-time information, beginning in the 1980s, and have since then invaded the world's commodity markets and complex financial derivatives in order to extract profits. They have created complex and non-transparent financial instruments that have enabled them to speculate profitably on the world's most valuable resources – food and energy. Nevertheless, new and emerging technologies will, in the near future, neutralize the speculative elements not only of those instruments but also of all other financial tools traded today – Stocks, Bonds, Currencies, Commodities, Debt Papers, including all of their derivatives such as Options and Futures.

These and other technological developments will also reveal the essential interconnectivity and interdependency in the global economy and commerce, where the insistence on buying a very low-priced T-shirt in a mall in Washington, D.C. will cause someone else on this planet to work for wages below the poverty level; or conversely, that the CO gases of a factory somewhere in India or China will make breathing the air even more unbearable, and some day impossible, say, in Northern California. Indeed, it is becoming increasingly obvious that our current financial, economic and commercial systems have exhausted their usefulness and the goodwill of most of the global and local participants. They have created an environment of global revolution and will soon foster new waves of terrorism and anarchy that will possibly set civilization back by decades, if no proper transition models can be found and implemented throughout the planet.

In this regard, we share the belief of a growing number of economists that we must move as quickly as possible from a speculative economy of financial pseudo values to a true value system, based on a fairly distributed, resource-and-consumption economy. We must concentrate on the innovative development of products and services that correspond to what humanity really needs and wants. Thus, productivity and development must be of the highest quality, driven by an educated and responsible consumer, and not the other way around, as it often happens today, when most corporate businesses direct and manipulate consumer behavior solely for their financial benefit, and when consumers become involuntary providers of asymmetric wealth for a few amongst us.

In other words, we need to counteract the negative economic effects of globalization, derivative-based finance, currency instabilities, and the interference of costly intermediaries (middlemen) that add little value to global commerce and only generate unnecessary expenses and obstructions of transaction efficiency in the process. One should not forget that middlemen are the creators of the financial industry of the past three centuries and are equally responsible for its impending collapse. Middlemen have absorbed 60 % of the world's total wealth without contributing in any way to human progress; instead, they have inspired wars and other highly risky behaviors, all for the sake of profiteering.

In the future, we must build wealth across the human population in its entirety, instead of concentrating it in the hands of a small minority. But we must accomplish this without falling into the traps of communism, socialism or corporate or crony capitalism, because all of these models have already failed us. On the other hand, we should not reject the elements of those systems that did work in the past, so we can build a more efficient world that can feed and protect all of the inhabitants of the planet in as fair a manner as possible. Thus, we need to return to sound commercial practices, based on the creation and exchange of real and tangible products and services and on a financial industry that provides solely fair credit transactions and account management.

Consequently, the future global economy will most likely move toward, or return to, a system that trades in actual products and services against the backdrop of other products and services, using one single economic unit of exchange. Within the future system of value-based commerce, a human being producing textiles in China will receive the same monetary reward as a person in Kansas City. And while prices in the market will fluctuate to some extent (as they do in any healthy economic system), we will no longer have a scenario whereby a man in Malaysia makes two dollars a day for producing a t-shirt (i.e. creating value), and a man in California makes $100 just for selling it.

In a well-functioning global economy, based on fair competition, comparable prices should be paid for comparable services with comparable quality, on a global scale. In this sense, the demise of the current financial industry may bring a newly found sense of justice and fairness to the economic performance of nations. Such a development is in line with the general trends of standardization of economic value and productivity, brought about through increased globalization. To speed up these evolutionary trends, we propose, in the present chapter, a fully automated global commercial and financial system that is designed from ground-up to be much better equipped to deal with the emergent global economy than the current financial and commercial industries are. Indeed, this system can play a major role in solving the current global economic crisis once and for all, without any social and political upheavals.

The Global Value Exchange (GVE)

The Global Value Exchange is built on the same QRT foundation that we introduced in the previous chapter. Even in the current, early stages of the development of electronic information and communication technologies, it is already possible to build a fully automated, global commercial transaction space. We would like to call this space the Global Value Exchange (GVE), because all of its operations and transactions are based on actual products and services, exchanged at their real value. The GVE system can operate under the mechanics of standard commercial systems but, more importantly, it can also function independently of the current global monetary system, such as today's global currencies and currency exchanges. The GVE system will be the first enabling technology that potentially unites all producers and buyers of any product and service, anywhere in the world, in one single, standardized, transaction space.

How Does the GVE System Work?

The global GVE transaction system is a generic market place, completely neutral vis-a-vis brands, producers, providers and buyers. It is implemented through a supercomputer-powered, QR-based, technology platform that electronically facilitates real-time *Business-to-Business* (**B2B**), *Business-to-Government* (**B2G**) and, in a separate technological space, *Business-to-Consumer* (**B2C**), *Consumer-to-Government* (**B2G**) and *Consumer-to-Consumer* (**C2C**) trading or exchange interactions for any product or any service between any global locations by any producer or provider. The GVE system also supports the function of automatically combining business or consumer groups either for composite business offerings, where several businesses compose a product offering to one consumption point (Consumer), or for bulk price negotiations, where a group of consumers can negotiate for bulk ownership of one expensive product such as an airplane, boat, or some time-share property, or where a group of consumers can negotiate a bulk purchase of a larger quantity for a better price. If we take into consideration the technological features of the QR-Search in the previous chapter, it should be clear how our QR-based technology could enable a system with such complex goals.

Of course, in reality no human must negotiate anything in the traditional sense, as the centralized supercomputer systems along with the global cloud extensions of the GVE will do it fully automatically and within milliseconds, anywhere, anytime, for anyone, on any scale, and with any transaction complexity.

Any entity that produces anything of value that can be verified, standardized and appraised, could participate in the GVE system. There is an equality of all participants, as the system no longer must clear trades that use currencies or other financial instruments having different arbitrary values at different times and places, and could thus deviate from the true value of the actual product they represent. In the GVE, one common currency instrument, which we have called the 'Global Economic Unit' (GEU) can be exchanged and banked on a real-value basis for those goods and services exchanged under the new system. The participants can always exit the system for legacy currencies, or can keep the GEU on account for further trading.

Figure 13 below presents a general model of the GVE electronic market:

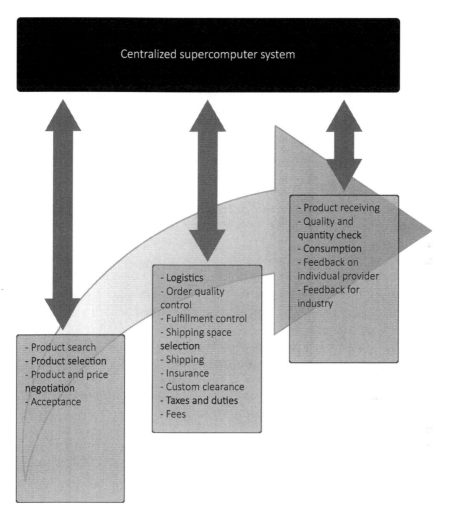

Figure 13. The GVE Market Model

The system will work much in the same way that a modern electronic stock exchange works today in order to settle trades. However, the functionality will be substantially more sophisticated, because it can, in order to complete a global trade transaction, automatically combine (through the "buyers' and sellers' selections") products and services across all of the necessary branches. For example, the GVE-executed purchase of a sea container-size lot of electronic equipment from Malaysia by a customer in Los Angeles could combine automatically, or by optimized selection (chosen, for instance, by best offer, or the safest way of shipment, or the speed of delivery), all services and dependencies needed to complete the transaction, such as container, shipping agent, vessel, clearing services, insurances and so on, without either seller or buyer ever having to contact any of those service providers.

In the GVE system, both the seller and the buyer can establish predefined sell-and-buy rules that the system will automatically implement for all affected transactions under the specified rules. These rules can be plain, conditional, or abstract in nature. In the case of product and service offers or requests where there are equally defined offerings, the buyer or the seller can set more selection criteria (real or abstract) to further narrow the transaction, or explicitly choose a transaction package that will close the loop between offer and purchase. Buyers and sellers will both have access to financing and leasing instruments that can be automatically combined as services to the transaction package. For example, a financial service can establish explicit rules and conditions under which it wants to assume the risk of financing. The GVE will then do the verification and, if all rules and conditions are met, will automatically establish the finance or leasing transaction. In this way, money, or any derivative of it, finds the customer automatically and fully efficiently, without any need of costly brokers or any other people in the middle. Simply making a resource available and determining the conditions of transacting it will suffice. The intelligent machines in the global cloud will do everything else.

The GVE system will automatically manage time frames for production and delivery, thereby differentiating the trade of actual, real-time inventory from what can be ordered, produced and delivered within, say, 30–60–90 days. This factor will enhance the planning practices for manufacturers and will also have a very stimulating effect on competition and efficiency.

The GVE architecture and process systems support the mediation of complex transactions involving large quantities of conditional rules and contingencies in milliseconds. It will reward producers of products and services equally, regardless of their geographic origins or points of consumption. Under the economic system of the GVE, the products of equal quality and specification will be valued equally, regardless of their origin, allowing a fair, globalized and competition-driven price equilibrium across the markets. It will thus stimulate the most economical creation of high-quality products and services instead of imposing differences by the instruments of the financial industry (unnecessary middlemen, agents, banks, currency inequalities, etc.). In an expanded stage, as represented in Figure 14 below, the system will not only facilitate business-to-business transactions, but will be extended into the government and end-consumer space as well:

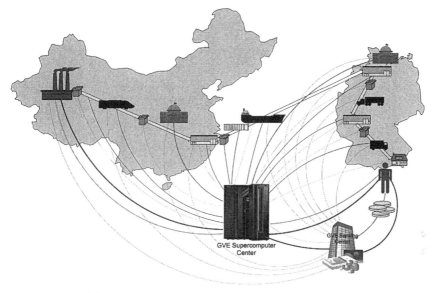

Figure 14. Expanded GVE System

How Does the GVE System Differ from the Current Business Transactions through the Internet?

As represented in the chart of Figure 15 below, the system automatically mediates all logistical and financial elements of a transaction from a neutral position, without a vested interested in either the buyer or the seller, and without seller and buyer ever having to contact or even know of each other.

Under the GVE technological solutions, a buyer in Germany could purchase a box of Asian fruits by being matched up with a selling farmer in Indonesia without ever having to know him, talk to him, or even visit his website. Nor will the farmer in Indonesia even need a website to sell his fruit to the buyer in Germany, as long as he has a registered supply-line into the GVE system from his nearest point of supply-line entry, preferably within reasonable distance of his place of activity.

Customers will ultimately be better-protected buying goods they never saw and will get better compensation in case goods or services are not the quality the customer thought he ordered, because the system can track transactions bi-directionally and maintain a quality-control feature. This feature consists of tracking, in all application spaces (B2B, B2G, B2C, etc.), a customer-satisfaction score, combined with a sophisticated feedback mechanism that will enforce far

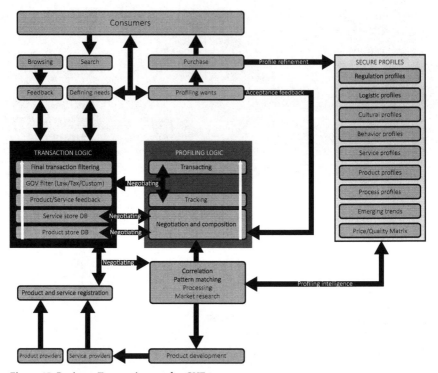

Figure 15. Business Transactions under GVE

better quality and service performance than are practiced under current systems. In other words, if you perform according to expectations, you will continue to get the business; if you perform poorly, the system will begin to ignore you. This quality control mechanism can track any specific cause, related to any aspect of transaction, including, producer, shipper, insurer, port handler, etc., and feed back individual performance grades.

As shown in Figure 16 below, the GVE system will compose mixed transactions (B2B and B2C, mixed products and services, multiple product origins or destinations) and execute all logistical and monetary requirements across all geographical and cultural spaces, in real-time transaction streams. It makes the most complex and most diversified business transactions a matter of complete automated execution logic within millisecond timeframes.

Any marketing-related expenses benefit neither the seller nor the buyer if the producer of products and services can complete the value chain without them. Such expenses will be driven to a minimum, because under the GVE system the seller can enter the supply line without ever having to print an advertisement or solicit a customer Likewise, the customer can simply define a purchase criterion under which the GVE system will select the most appropriate product or service

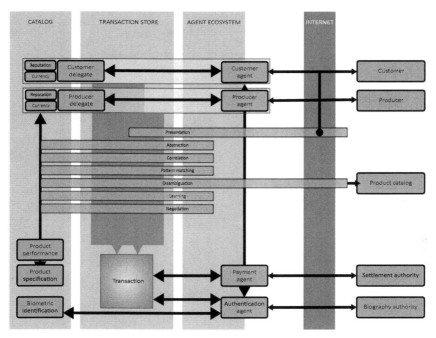

Figure 16. GVE Transaction Streams

offering that fulfills his or her needs. Often it is not even necessary for the buyer or consumer to make an explicit demand, because the system will use sophisticated profiling technologies to anticipate future demand or needs, and compose appropriate product and service offerings (let us remind the reader that the Q-Search engine described in Chapter 5 is an integral part of the GVE system). Naturally, such offerings are not restricted to physical products, as they can also be some element of information, education or helpful guidance. Under the GVE model, all of those offerings (products, services, or information) are processed in the same logic engines of the system.

Furthermore, the system will, automatically and based on the global sum of all buyers, give feedback to the creators of products and services on how to optimize these, so that they become better adapted to the market place. Under the proposed GVE system, the marketing expenses are mostly invested into product adaption, which benefits the producer and the buyer, instead of going toward trying to convince a buyer to obtain a product or service that is suboptimal for his or her requirements. This system will significantly fuel innovation and streamline production, as seen in the chart of Figure 17 below, well exceeding the current economic models of supply and demand.

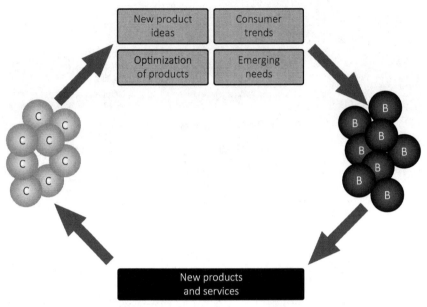

Figure 17. Streamlining in the GVE System

How Will the GVE System Transform Electronic Markets?

The GVE system is designed to enable participants in electronic markets to engage as informed, direct partners in transparent transactions. For this purpose, it incorporates several specific mechanisms that provide an information flow and control relations much better than proprietary markets. In this section, we shall first describe the behavior of current, electronic proprietary markets, including the components essential to them in a *B2C environment*, with the actions they permit or deny their participants. Then we shall contrast them with the mechanisms that the GVE introduces in order to remove such proprietary barriers. We use the B2C environment here as premier example, because it best illuminates the problems the GVE solves, but the system's functionality scales to all of the other spaces (B2B, B2G, C2C, C2G, etc.).

Electronic marketplaces today concentrate processing on centralized servers under the control of the market, as shown in Figure 18, below:

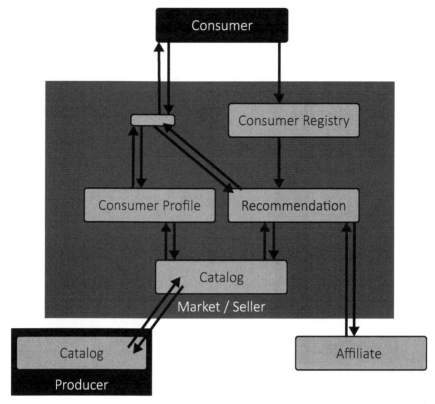

Figure 18. Current Electronic Market Model

A buyer's interface – customarily a web browser – requests information about products and services from the market. The buyers express a search request in terms of products/info contained in the market's central product and service catalogues. The market's role is to furnish product information to the buyers, based on their immediate request and their past behavior – their identity and previous online requests and purchases. The response will include information immediate to the request. The response may also recommend secondary products to the buyers, based on their profile and history.

The market may permit buyers to arrange their interface to display selected information streams that it delivers and it may also allow them to use one-off search "agents" to define what they are looking to procure. Despite such extensions, consumers and producers communicate through the market space only. That space is dominated by the markets' own interests, in combination with those of select sellers who have invested to create the market space or pay to participate in it. There is very little opportunity for buyers to have direct interaction with the full range of existing sellers.

Given the markets' dominant position, they have been able to develop and deploy sophisticated techniques to profile buyers and push towards them the products they want to sell. Markets increasingly rely on profiling technology, which they use to present products to clients. Each time the buyer searches for products and services to consider, or purchases a product or service, he or she reveals more about his or her personal profile and tendencies. Markets observe, record and analyze this behavior, using semantic analysis and agents, and their profiling technology adjusts the information stream accordingly. This process results in a certain manipulation of the buyer on the part of the market, because the buyer has no direct access to the real offerings of goods and services outside what the Seller in the market is inclined to show them (see chart in Figure 19 below). As effective as these techniques may become, they are bound to act in the market's interest only.

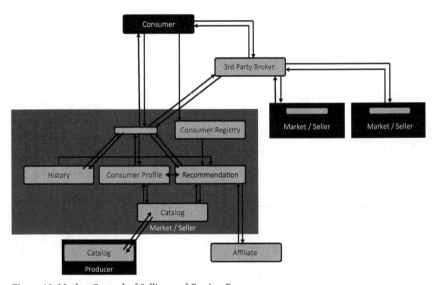

Figure 19. Market Control of Selling and Buying Processes

Buyers, on the other hand, are restricted from expressing directly to producers what it is they want to buy, because there is no functionality available to permit them to communicate directly with producers, or to dynamically and continuously refine and request products and services. Buyers have very limited means to independently profile products and services without the aid of the market and the seller. While buyers can resort to services, which provide similar technology to search for products based on specifications, the capacity is limited to one-off use.

A structure where agents and semantic analysis can work dynamically and continuously for the buyer would greatly improve the bidirectional flow of commercial information and enhance the flow of goods and services of real value, as represented in the chart of Figure 20 below:

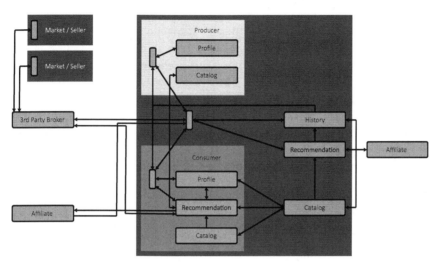

Figure 20. The GVE Open Market Model

As this chart indicates, a shift to an open market paradigm will yield control of these functions to the consumer and empower the buyer to independently seek goods and services that are available without limitation to what the selling market dictates. It will also help producers provide goods and services that reflect the buyers' true needs. This would give producers a shorter, more efficient route to turn over sales without having to guess what to produce in order to fulfill the needs of consumers. Many costly scenarios could be avoided, such as warehousing bottlenecks, middlemen who tack on extra cost, but deliver no added value in the supply chain, and expensive marketing and advertising campaigns, aimed at creating false needs for products and services.

We have designed the GVE system to operate this paradigm shift and remove such limitations. By reducing the market's control over information access and flow, the GVE facilitates transparent, symmetrical business relations between market, buyer, and seller. Before we show how the GVE system accomplishes this goal, let us describe the essential components of the current proprietary markets and how they need to be reconfigured.

Several researchers have already shed light on the nature of markets in general and electronic markets in particular. For example, Guttmann (1998) describes the cognitive processes whereby the individual participants, in order to achieve their commercial goals, obtain, evaluate, and act on the information they can gather about the market, its products and the other participants. In turn, Porter (2001) analyzes businesses from a similar perspective, showing how the processing of information flow in business activities is analogous to the structure of an individual's cognitive processes.

According to these two researchers, all products and transactions are manifest in electronic commercial processing of information as well. This means that, the essential structure of commercial processes concerns the flow of information, where it originates, how it is located, who controls its availability and dissemination, how participants evaluate it, and where it ends up. The participant who controls the content and flow of information necessarily dominates the market.

Both researchers identify cognitive/business phases through which the participants pull, push, drive, collect information as they decide what and with whom to buy and sell products. The described processes are, in principle, neutral: information flows and contributes to decisions. The concrete manifestations, on the other hand, constitute a very different world. When we project the abstract model onto concrete market instances, we soon discover that, instead of a neutral world, the e-market ecosystem is one in which the markets control the information flow, active roles are limited to market participants and/ or their delegates, while the buyers' role, or even that of the independent seller, is limited to passive interactions. This imbalance extends even to emerging mechanisms that use a buyer's history ostensibly to promote his ability to find suitable products, but actually benefit the market itself by reducing its merchandizing costs.

Guttmann proposes a business process that comprises seven phases:

1) Need establishment; 2) Product information; 3) Merchant information; 4) Negotiation; 5) Contract; 6) Delivery/fulfillment; and 7) Service. Porter, in turn, outlines the business processes for a commercial entity: 1) Acquisition; 2) Design; 3) Marketing; 4) Sales; 5) Delivery; 6) Service. In these terms, we can depict the processes in an electronic marketplace as a network of relations among presentation, processing, and storage components. The chart in Figure 21 below demonstrates the current electronic market's dominant role:

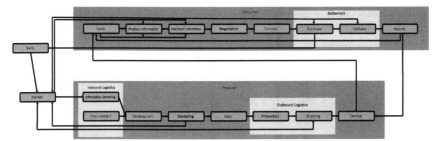

Figure 21. Dominant Role of Current Electronic Markets

In this chart, the only direct connection between consumer and producer is the post-sales link for warranty service.

Another indication of the imbalance is a simple accounting of the allocation of components to participants, on the example of Amazon.com, in Figure 22 below:

		Need	Product Info	Merchant Info	Negotiation	Contract	Delivery	Service
Market			Navigation, market catalog, category u/i, listing u/i, product u/i			Invoice u/i	Shipping u/i	Support u/i, rating u/i
	Suggestion agent		Inventory api, buyer history, market widgets				Shipping delegate	
Seller			Seller widget					
Buyer			Registries					

Figure 22. Components of Amazon.com

By contrast, the GVE system handles the market and consumer interactions in a fundamentally different way. All of the current market systems have been created primarily to serve the "middleman." In fact, these systems are middlemen themselves. Ironically, in the current global market, middlemen contribute the least, risk the least, but get rewarded the most. The GVE system aims to re-structure the middleman as an automatic global technical facilitator that lets the producer of products and services and the consumer control all of the trans-action processes. To explain how the GVE system will accomplish this objective, let us examine the phases identified by Guttmann and Porter in relation to existing electronic markets, then contrast the observed technology with the GVE needs and describe the alternative technology these needs entail.

1) *Need establishment phase*

In most e-markets, the products are "just there." Although the market is "electronic," the paradigm is that of the physical department store. In an online shop, a product catalog determines the products presented. In an auction house, this restriction is somewhat relaxed, as the sellers place the items on sale, but the market still dictates the terms and the site organization. Little is done to establish a direct connection between a buyer's (or seller's) condition and its manifestation as a "need" (or a "surplus"). Commodity markets do provide some support for need determination. For example, e-pharmacies can track prescription history and refill prescriptions proactively, while replacement supply markets make similar arrangements for products like paper or ink.

Current e-markets are deficient, however, insofar as the market itself controls these processes and tailors them to its product offerings. The buyer must make privileged information available to the market and must acquiesce to the market's determination of imminence and suitability. There are only a few sites that allow offered items to be correlated to an independent catalog rather than to be submitted strictly in terms of a market's controlled vocabulary. It is impossible to offer items either directly, for example, by scanned product serial number, or free-form text. This serves to frustrate mechanisms that act in the other participants' interest, such as offering directly from another participant's inventory, or establishing a need by implicit analysis as a side effect of other buyer/seller information flows. By contrast, GVE will implement a need analysis mechanism in the form of autonomous buyer/seller electronic delegates that are either tasked with specific observations, or will continuously and discreetly observe the participants' actions both in-market and at-large, suggesting appropriate initiatives. This means that a service agent that tracks the status of the buyer's equipment is under the buyer's rather than the market's control and, in the event of a need, engages an electronic information agent to interact with several suppliers to locate products on the buyer's behalf.

2) *Product information phase*

The successful e-marketplace sites, such as Amazon, demonstrate remarkable acuity, ingenuity, and engineering capacity for extracting information from a buyer's actions on its site and using that background information to optimize dynamic information flow to the buyer. The results are impressive, and the principle finds wide acceptance. These sites go so far as to make the information available to third-party participants for use in affiliation with them. They even drive spin-offs, which factor the technology out of a given site and apply it in a generic form to any given market site. As effective as this effort is, it falls short of GVE goals, because its decisions are based on a unidirectional information flow from buyer to market, which serves only to increase the purchase volume per time-x-information. The information is controlled in all aspects solely by the market.

By contrast, GVE realizes similar proactive CRM technology, but in a form in which the buyer/seller participants exercise control and in which information passes not only from buyers to the market, but also from product and seller performance history to buyers, so that the latter can evaluate offerings. In other words, the participants involved, rather than the market, record their own histories and have recourse to the histories of other entities, can formulate the criteria for their use and evaluate the results. One consequence of the shift from market to participant control is that the current static agent technology must yield to mobile agent infrastructures in order to permit computational components that act in a buyer's or seller's interest to migrate among markets while fulfilling security and authentication requirements.

3) *Merchant selection phase*

In current e-markets, the merchant is always in complete control. There are some sites that offer multi-market product information, but their scope is limited and the collected merchant background information is limited to subjective reviews. Changes to access technology are going to transform aspect of e-markets in the near future. A shift from workstation interfaces to handsets (as with mobile telephones) will force the merchants to yield control over the information flow initially to a carrier, but ultimately to the buyer/seller participants. One reason is the participants' more intimate relation to the device. The other is the limited physical real estate available in the devices, which compels each offering to coexist with others rather than to command control over the entire interface.

At present even an electronic bazaar like EBay dictates not just the content, but also the avenues to follow in order to reach this content. Some sites do attempt to appear neutral. For example, in the case of Amazon, they integrate offerings beyond their static stock from affiliates, from third-party sellers, and even from competing sites. This appears to act in the buyer's interest, but all control flows through the market with the goal of increasing the buyer's awareness of and facilitating sales from their stock.

The GVE will reverse this relation to support both the buyer's and the seller's interests by providing them with means to construct their own venue, dictate the terms of its use, and invite offerings from their counterparts. This evolutionary process is just starting to manifest in the form of self-configurable gateways, such as netvibes. As devices shrink and "ecosystem"-based technologies allow ever more access, pressure will increase to relinquish control to the participants.

4) *Value chain construction*

   The control shifts that GVE introduces into the product information and merchant selection phases will ultimately expose the value chain, rendering it mutable. The current commercial transactions entail disjointed, producer and buyer's business processes revolving around predefined products and intermediate market staging. By contrast, the GVE model will employ agents that couple the information in the producer's design and production phases with the information in the buyer's product and merchant selection phases, so as to permit a buyer to compose a specific value chain from production to transport to retail intermediaries to delivery for a specific need.

5) *Negotiation*

   The dominant e-market business model prescribes both prices and product specifications. The market may publicize reduced-price offers and it may suggest a wide spectrum of products, but, in fact, the buyer has limited control over price/performance trade-offs, because they have almost no control over detailed product specifications. The buyer's options are to "take it or leave it." The only limited exception is auction sites, but even in these cases, the actual products are immutable. Since the GVE enables the coupling of the buyer's and the producer's information and design phases by allowing the buyer to selectively compose the value chain, it puts them in a position where a buyer – or its GVE projected electronic agent within the protected user profile of the system – can negotiate with producers about trade-offs between design, performance, time, and price.

   Furthermore, the GVE system proposes to replace the fixed-price transaction form with the bidirectional continuous auction, prevalent in financial markets. This means that prices follow supply/demand while price history and performance are available for buyer/seller agents to incorporate in decisions.

6) *Contract construction phase*

   Recent market sites, such as Doba, demonstrate how to extend the market beyond the warehoused product to include the shipping process. GVE will take the next step, where the user will control the supply chain construction beyond selecting only from the market's chosen suppliers. It should be possible for the buyer to issue an open request, to be satisfied by suppliers who can perform the service from manufacturing to consolidation to delivery.

7) *Supply-chain construction*

   In a manner similar to the changes that GVE will introduce to the value chain, it will also facilitate purpose-specific, supply-chain creation.

8) *Automated composition of negotiation, contract, and settlement phases*
The GVE implements, under the principles of the supply-line creation, additional functions from the perspective of the consumer:
- C2C product trades are instant, straight transactions and auctions between consumers, with the added benefit of completing the transactions, including system-constructed shipping and payment methods, without any participants needing to solve those problems themselves, as it is necessary on EBay, for example.
- C2B bulk purchase transactions are automated, whereby clients need only indicate their intent to participate in a buyer group, without having to know or meet the other clients: the GVE system will serve as the authenticating authority and negotiate automatically with the supplier on behalf of the consumer group.
- C2B consortium purchase of a single product of great value that is owned and shared by multiple Clients, such as a vacation home, an expensive machine, a boat, or an airplane, is also automated. The GVE will facilitate this process by bringing together profiled Clients and making such expensive consumer items more affordable as joint-purchase property, under various legal implementation options that best fit specific product and client profiles.

9) *Settlement/delivery/fulfillment phase*
Most e-markets act just as an authentication intermediary for payment and delivery processing, leaving the completion of the transaction to external delegates. Barter sites introduce the significant feature in which the market serves as the bank for payment settlement as well as the intermediary. The benefit is that since the market controls the currency, it sets the currency's worth in two ways: The first is through the exchange rate to external payment instruments. All known sites prescribe a 1–1 exchange of internal value for external and limit the worth by prohibiting the reverse exchange. The second way is through the length of time permitted for settlement. The site can require immediate settlement, or it can permit settlement to be deferred so that an internal debt can be carried over some interval before either a trade or an external payment must cover it.
GVE will strictly adhere to its transparency policy and allow the internal market to determine both the exchange and the inflation rates.

10) *Service phase*
Numerous e-markets offer means to author and retrieve product reviews. The sites foster internal, community behavior and attempt to connect to external social networks in order to promote customer bonding by serving as a focus for these activities. The deficiency here is that the markets bind the information to their realm, determining the subjects and controlling the

access and the terms. The subjects are limited to products available through the market. In most cases, access is limited to registered participants. In all cases, reviews are limited to idiosyncratic discursive contributions.

As an alternative, GVE will permit each participant (whether buyer or seller) to carry his or her evaluation from one market to another and to participate on the basis of third-party authentication, in addition to local registration. Furthermore, by augmenting discursive information with automatically collected performance and service information, GVE will be able to offer objective background data about product value.

Taken as a whole, these mechanisms will shift control over electronic components from the market to buyer and seller. Figures 23, 24, and 25 below represent charts that show the net advantages of the GVE system over the current electronic markets in terms of their components:

Market Components	Need	Product Info	Merchant Info	Negotiation	Contract	Delivery	Service
Current		Navigation, market catalog, category u/i, listing u/i, product u/i			Invoice u/i	Shipping u/i	Support u/i, rating u/i
	Suggestion agent	Inventory api, buyer history, market widgets				Shipping delegate	
GVE		Market Catalog, category u/i, listing u/i, product u/i	Market Catalog, category u/i, listing u/i, merchant u/i	Bidding u/i		Shipping u/i, tracing u/i	Rating u/i, tracing u/i, support u/i, service u/i
		Inventory api	Rating system	Multi parameter, bidirection al, continuos auctions		Shipping delegate	

Figure 23. Market Components in Current Electronic Markets and GVE

Seller Components	Need	Product Info	Merchant Info	Negotiation	Contract	Delivery	Service
Current		Listing u/i					Support u/i, rating u/i
		Seller widget					
GVE		Crawler, solicitor		Bidding u/i	Contract u/i	Tracing u/i	Tracing u/i
		Seller widget	Rating agent	Contract agent, auction agent	Contract agent	Supply-chain agent	

Figure 24. Seller Components in Current Electronic Markets and in GVE

Buyer Components	Need	Product Info	Merchant Info	Negotiation	Contract	Delivery	Service
Current							Tracing u/i, rating u/i
	Suggestion agent	Registries, product agent		Bidding agent			
GEE		Buyer catalog, product u/i, review widget, widget sandbox	Performance index	·	Contract u/i	Tracing u/i	Tracing u/i
	Buyer inventory api, prescription list, service agent	Buyer history, product agent, solicitor, registries, rating agent	Merchant agent, solicitor, buyer widget	Contract agent, bidding agent	Contract agent	Supply-chain agent	Service agent

Figure 25. Buyer Components in Current Electronic Markets and in GVE

Proactive, protective mechanisms

The GVE system supports the fundamental principle that any business must provide value-adding product and services to its customers within the appropriate social norms and institutions; ideally, it should also contribute to the benefit of society. This principle obviously applies to banking and finance too, as profit will be a natural outcome of the business activity. But what is much more important, the GVE-enabled flow of steady profits will bring stability to business supply chains. Clearly, under the current system it is possible to have huge profits in the short term by deviating from such steady value-adding practice. Yet, this predatory practice will not last and will often end up getting the business into instability and ultimately into bankruptcy. (Heng et alia, 2009)

The GVE solutions will eliminate unnecessary legions of middlemen, profit-reducing intermediaries (financial practices and instruments of pseudo value), including interest. The GVE system will not automatically mediate a transaction unless it produces a simple, built-in profit for the creators of products or provider of services. Naturally, the supplier will be able to override this feature of the system, but the GVE will calculate out the losses the transaction will produce if overridden.

The GVE system will also minimize the elements of hedging and risk. It is also transparent in nature, promoting information disclosure. It will establish a proper framework to consider the appropriate level of disclosure so as not to result in unnecessary overflow of information or a loose, under-informing regime of disclosure. (Mohd and Engku 2008, 107).

The GEU (Global Economic Unit): Its Functionality and Added Value as a Single Currency

The means of payment and a measure of value (pricing unit) are integral components of modern Web-based trading platforms. Besides private currency and direct clearing of traders' credits and debits, most trading platforms use institutionally backed currency as a measure of value. It is not new that institutionally backed currency is a politically backed unit, the value of which is dependent on the policies of the respective government or central bank issuer. It creates a system burdened by the effects of exchanging currency between issuers, such as paying exchange fees and coping with the uncertainty of fluctuating exchange rates for possible future transactions. Because the measure of value as a political unit is very unstable, traders suffer losses and inequity ensues among participants.

We propose to use the Global Economic Unit (GEU), which is a universal measure of non-political value, providing a unit of account, independent of all national currency units. The GEU as an electronic currency unit used inside the GVE can be exchanged and banked on a real-value basis for those goods and services exchanged under the GVE system. In this scenario, all intermediates such as banks and third-party money services are unnecessary. GEU as a measure of value will remain stable and become a unit of productivity.

Trading participants will, however, have the ability to choose which pricing unit to use to settle trades under the GVE payment system:

The GVE proposed measure of value
The GEU institutionally backed currency

Of course, the system will support transactions backed up by currency in the traditional manner but, over time, participants will likely discover the full benefits of trading with the GEU.

The GVE System and Taxation

The GVE participants should treat the GEU as any other legacy currency in terms of taxation. Trade, barter, and other transactions are assessable and deductible for income tax purposes to the same extent as other cash or credit transactions. Therefore, the advantages and disadvantages are similar to ordinary cash revenue or expenses. Trading may result in tax liability when selling and may attract tax credit when purchasing. It may involve fringe benefits, sales tax, and capital gains legislation – depending on the domestic country's taxation policies. GVE is not designed to be misused as a means of tax avoidance or evasion.

Trade Examples within the GVE system

A traditional trade example (legacy currency to legacy currency): A person from London wants to buy commodities to produce bricks. He posts a buy-request for clay on the GVE platform, where he has been offered best match in terms of price and logistics. The GVE system negotiates contract details, arranges an agreement and executes the trade, by exchanging British pounds for clay, all of this being done automatically.

A GEU trade example (GEU = GEU): Trade details are the same, except clay is exchanged for the price in GEU's, not legacy currency.

A barter trade example (quoted in part from Bartercard): A restaurant wishes to "barter" for $10,000 worth of printing and advertising using meals and drinks. No printer would ever want $10,000 in meals from the one restaurant. However, by selecting a printer who participates in the GVE, the restaurateur pays the printer with 10,000 GEU's (for the purpose of this example, 1 GEU would equal 1 USD). The printer's account is "credited" and the restaurateur's account "debited," so the restaurant now owes 10,000 GEU's in meals to the GVE network, NOT the printer. This is repaid as various GVE members come to the restaurant (new business) over the following months. Hence, the restaurant has paid for its printing requirements with meals, using an interest-free "line of credit" and at a cost of approximately 30 cents per dollar (approximate profit margin), in replacement goods cost. In fact, there will be no opposition from governments to such transactions, because taxation is easily tracked and applied just as in traditional transactions.

Indeed, this taxation clearing is much more convenient for all parties involved. The printer can now use the GEU's earned to buy office furniture, advertising, car repairs, stationery, courier services, color separations, accounting and legal services, even a family holiday. These are all commodities for which the printer would normally pay cash. By using his GEU's, the printer conserves his cash and benefits from the extra business.

A cross-currency trade example (GEU's = Legacy or other currency): A buyer from Romania wants to buy parts for his car. As an unregistered user, he decides to browse GVE for offers by sharing his wish list with the system. He has been matched by a seller who only accepts GEU's as a settlement currency. The buyer interacts with the GVE's banking and currency issuing system, to buy GEU's. They are fixed value, backed by productivity/purchasing power/real value services/products/commodities. The buyer exchanges his legacy currency to GEU's in order to acquire goods over the GVE.

Cash-out or GEU/legacy currency exchange: If a user wants to exchange GEU's for any other legacy currency, there are several ways to do that:
- The system finds another counter party and interacts with a third-person who wants to acquire GEU's (because in the starting phase, most of the GEU's are created over the barter exchange trade model).
- The GEU holder simply exchanges to legacy currency via settlement with the bank interface, which is also a part of GVE.

The Trade Value of the GEU

The value of the GEU is set to the value of a basket of legacy currencies and other units of value, calculated in terms of basic commodities and labor. This basket will not change except under extraordinary circumstances (the collapse of a major national currency such as the US dollar, for example) and will become a permanent definition of the GEU. The currency value is therefore stable, easily adjustable, and predictable.

The Security of the GVE system

The inbuilt security features of the GVE system go beyond those of any presently existing Web-based system. It makes the GVE immune to cyber attacks as well as to fraud and outside manipulation. Fraudulent transactions are automatically eliminated from and penalized by the system. As already discussed, transactions are anonymous, so that the identity of the trading partner is completely protected and can be disclosed only if the disclosure is mutually desired and agreed on by all of the partners involved.

Technological and Scientific Requirements for the GVE System

To drive this project of enormous computational proportions, one needs to use the power of the latest Enhanced-Intelligence tools such as the QRT platform (presented in Chapter 5 above). One also needs to assemble a transdisciplinary and intercultural team of specialists in problem-solving, including computer scientists, physicists, mathematicians, linguists, economists, humanists, psychologists, and other social scientists, who work across disciplines and cultures to implement the GVE system. In fact, today there are a number of privately owned and operated Supercomputer Centers that are perfectly equipped to drive the GVE system in its fully developed architecture. There is also sufficient scientific and technological know-how in the world to make it possible, for the first time in the history of humankind, to implement massive and complex global projects such as the GVE system.

Implementation Time Frame and Cost of the GVE System

We have calculated that, given the appropriate QR technology and the right team of scientists, we can build a basic GVE system and make it fully operational for at least a particular industry segment (Food, Automotive, etc.), within twenty-four to thirty months, at the very reasonable cost of about $60 million. This modest investment, which is small change for any global venture capital investment firm, could be recuperated in a matter of months and could bring huge revenues through the volume of nominal fees charged for each of the millions of daily transactions on the system. The system could then grow to its full global potential by being financed entirely through revenue.

Every time the GVE system completes a transaction between the participants, it has the potential to charge a fee. Such fees will in general be very small, typically between €0.001 and €0.05. Many transactions will be free, such as the listing and simple searching of products and services. Other, more complex, transactions will require a subscription, such as the ability of a potential buyer to execute some complex search every minute, to "fish" for an acceptably low price.

Although the price charged for any single transaction is very small, the system is designed to execute hundreds or thousands of transactions every second, and there are 86,400 seconds in every day. Thus, the potential revenues are exceedingly high. In addition, the system will provide extended features (limit orders, for example) to accounts that have paid subscription fees, so that prices for subscriptions to extend the automated abilities of the bargaining entities will range from a few euros to potentially thousands of euros a month.

All prices charged by the GVE are themselves a form of economic friction, and therefore increasing fees will have a direct and negative impact on transaction volume. In addition, the GVE itself is the reservoir of GEU (Global Economic Units), and it can, therefore, regulate the supply as required, by making more GEU available within the system, or by removing GEU from the system. As already mentioned, the value of each GEU will be fixed by a basket of legacy currencies and other units of value; in the interest of financial stability, that basket should not be changed except rarely and in response to significant events (the collapse of a major national currency, for example).

Conclusion: Evolutionary and Ethical Implications of the GVE System

The present conditions in global finance and trade have clearly shown the inadequacy of economic practices in a global environment and the need for the kind of economic reform that would remove unfair ways of doing business that privilege certain participants over others in the global economy.

Furthermore, in a global reference frame, obsolete distinctions such as that between "developed" and "developing" countries lose their meaning and reveal themselves as mere conceptual tricks, designed to perpetuate the old financial and economic inequities. The law of human evolution itself renders such distinctions inoperative. From the point of view of human evolution, there are no human beings or societies that are fully "evolved" and no economic or social system that has reached "perfection." In this regard, all countries are developing countries, and all of them can bring valuable contributions to the global community and its economy, which must be based on principles of equity and fairness for all the participants. The Global Value Exchange solution that we propose is, and must be, fully grounded in these principles.

In conclusion, the main practical goals of the GVE are: 1) to reintroduce, in the emergent global economy, the fair exchange of actual goods and services of real value; 2) to create stability through the precise calibration of economic units and to remove different currencies or instruments that have different values in different places; 3) to remove the highly speculative edge and unnecessary intermediaries that charge exorbitant interest, leverage between creditors and debtors, and are generally oriented toward usurious practices. In this sense, the GVE automatically adopts and applies some of the basic principles of traditional commerce, based on fair and mutually advantageous trading and honestly earned profit. What is different, however, is that, for the first time in human history, these fundamental principles will be inbuilt, automatic features of the global economic system, based on a truly free market, without government or any other outside intervention, and impervious to fraud, corruption, and abusive manipulation.

Thus, one could argue that the GVE system is a result of a long evolutionary process and constitutes a new stage in human development, viewed within a global reference frame. Within the GVE system, the creation of products and services will become a co-evolutionary process, governed by natural feedback loops between producers and consumers. This evolutionary principle, in full accord with the Quantum Relations Principle, will apply always, whether in business-to-business, business-to-consumer, business-to-government, or consumer-to-government relationships.

The GVE system will therefore generate a symbiosis between those who produce actual goods and services and those who ultimately consume them. With the help of this system, not only commercial transactions between producers and consumers, but also all other human interactions will be symbiotic and based on mutual benefit, rather than on greed, exploitation, and enslavement. While the GVE system will help prevent flight of capital, tax evasion, market protectionism, and abusive economic and financial practices in general, it will also help eliminate social conflict, including labor union and management disputes. It will naturally promote lucrative and wise investments of local and global capital, based on the real value of goods and services, as well as social responsibility and solidarity throughout the world.

Chapter 7.
The Global Healthcare Exchange: Managing Human Services in the 21st Century

Healthcare is one of the most important and challenging problems confronting the world today. Anyone can observe that current healthcare systems operate at a level far below potential. Even if healthcare is not on the brink of an actual collapse, as some experts believe, its current problems threaten to worsen over the coming years. Indeed, the accelerating pace of innovations such as new medicines, new diagnostic and treatment techniques, the molecular genetics revolution, as well as many other innovative tools, are widely surpassing our current abilities to practically implement these techniques in a wide and equitable fashion, both locally and globally.

Many healthcare systems in industrialized countries have high-tech medicine, but must continually fight with the pervading disequilibrium of costs and levels of quality for the citizens at large. Very often, attempts to solve these problems lead to restrictions of healthcare services. These efforts frequently create an additional loss of efficiency by expanding the costs of administration overhead. Although the quality criteria of healthcare systems are not easy to determine, and the World Health Organization's ranking system has shortcomings, ranking healthcare quality versus spending can prove the discrepancy: the US, for example, can be found within the five biggest per capita healthcare spenders, but the healthcare quality is only around the 30th rank. (Cited in http:// www.buzzle.com/articles/best-health-care-in-the-world.html)

The breakdown of healthcare spending in the United States in 2010 is as follows: "hospital care (about $788.9 billion), physician and clinical services ($535.8 billion), prescription drugs ($260.1 billion), nursing home and home health ($226.4 billion), dental care ($107.9 billion) and other items totaling $681.1 billion. Registered hospitals totaled 5,815 properties in 2009, containing 951,045 beds serving 37.5 million admitted patients during the year." On the other hand, "only a relatively modest amount of money is spent on preventive medicine and health education," with "about 70 % of healthcare funds" being spent on chronic disease. (Cited in http://www.plunkettresearch.com/ health%20care%20medical%20market%20research/industry%20overview) A re-

cent Thomson Reuters report found that the U.S. healthcare system wastes between $505 billion and $850 billion every year, with around 22 percent of that
going on fraudulent insurance, kickbacks for referrals for unnecessary services
and other scams. (Cited in www.reuters.com/article/2010/01/18/us-healthcare-
fraud-idUSTRE60H01620100118)

Needless to say, the ratio of spending versus efficiency is problematic on a
global scale. The situation in Europe, for example, is hardly different. The
medium increase of healthcare cost is far greater than the rise of gross domestic product (GDP) and it goes invariably to the taxpayer. According to
Reuters, "some 180 billion Euros is lost globally every year to fraud and error
in healthcare – enough to quadruple the World Health Organisation's and
UNICEF's budgets and control malaria in Africa." (Cited in www.reuters.com/
article/2010/01/18/us-healthcare-fraud-idUSTRE60H01620100118) In turn, a
study by the European Healthcare Fraud and Corruption Network (EHFCN)
and the Center for Counter Fraud Services (CCFS) at Britain's Portsmouth
University found that 5.59 percent of annual global health spending is lost to
mistakes or corruption.

According to the World Health Organization (WHO), the "cost of healthcare
is straining governments and individuals around the globe" with 100 million
people a year being pushed into poverty by their medical bills. For governments,
the challenges of funding healthcare are increasing because of aging populations,
a growing burden of chronic diseases, and the introduction of more expensive
treatments." (Cited in http://www.burrillreport.com/article-a_global_health
care_crisis.html)

The expense of developing new medicines, increased by complex and lengthy
regulatory requirements, contribute heavily to healthcare costs. In the United
States, the average cost of bringing a new medicine to market is estimated (as of
2000) at $802 million. (Cited in http://csdd.tufts.edu/) The genomic revolution
promises to increase the number of potential medicines by two orders of
magnitude, so costs will increase accordingly. Costs of clinical trials are also
growing fast, partly related to increased difficulty of patient recruitment.

Thus, the cost of healthcare is enormous and far greater than can be justified
by the quality of services delivered. The impact on the individual citizen is
devastating. Again in the United States, the statistics are shocking: uninsured
adults are 25 percent more likely to die prematurely than adults who do have
insurance; patients who have serious health conditions, such as heart disease or
cancer, face risks of premature death that are 40 percent to 50 percent higher;
about half the families that file for bankruptcy do so, at least in part, because of
healthcare debt; half of home foreclosures result, at least in part, from health care
debt; the administration overhead devours at least one quarter of the total
medical spending.

Preventable medical errors and uneven quality are also main areas of concern. A typical Institute of Medicine (IOM) report found that "between 44,000 and 98,000 Americans die each year from medical errors. At least half of these are preventable. Many more die or have permanent disability because of inappropriate treatments, mistreatments, or missed treatments in ambulatory settings." (Cited in http://books.nap.edu/books/0309068371/html/) The annual US cost of medical errors is estimated to be between $17 and $29 billion. The reported reasons for these errors are related to the breakdown in the communication structure of today's healthcare systems. Staff-related problems (quality of provider–patient interaction, morale, provider communication) are another major factor in poor healthcare quality. (See http://www.gallup.com/gptb/) Growing emphasis on formal standards of care might alleviate some problems related to poor and inconsistent care. However, there is now a proliferation of competing standards. Clinicians often complain that standards are drafted with an aim of reducing marginal costs, rather than promoting quality of care on an individual basis.

Another serious issue is the inequitable provision of services. For example, large variations exist in the quality of health insurance plans. Many plans are restrictive. Health Maintenance Organization (HMO) plans involve significant delays as medical boards must review and authorize even routine services beforehand. Members of lower-quality plans are more likely to receive poor treatment and have higher mortality in response to treatment (Erickson et alia 2000). The huge inefficiency in most healthcare systems worldwide makes it very difficult to avoid the daily situation of "luxury medicine for the rich – minimal care for the poor." More and more restrictions of services for basic healthcare can be observed, but most emergency actions targeted at keeping the system alive only succeed in producing more administration overhead. On a global scale, the situation is devastating. For great sections of the global community a consistent healthcare system is simply nonexistent.

Finally, healthcare fraud is a serious problem with global consequences. The loss of resources that could be utilized to improve healthcare is far greater than perceived by the general public and healthcare administrations. For example, the US National Health Care Anti-Fraud Association (NHCAA) makes conservative estimates that 3 % of all health care spending ($68 billion) is lost to health care fraud alone. In turn, the potential losses of resources for healthcare across Europe caused by fraud and corruption are estimated to be at least 30 billion Euros each year, and may be as high as 100 billion.

Some of the more obvious reasons for fraud include personal greed on the part of incompetent or unethical healthcare providers, or patients trying to elude restrictions of services. Generally, there is no consistent control over which services have been actually rendered, leaving the insurer with a great risk of financial loss by false claims. This situation is especially out of control when the patient is not directly integrated in the billing process. Moreover, more and more methods are being devised to "milk" the public healthcare system. An even worse situation is that of a patient put at risk by a physician who uses unnecessary procedures with potentially severe side effects only to drive up the patient's bill. Consequently, fraud contributes to a spiral of more and more restrictions to public healthcare systems, as well as to an ever-rising need for out-of-pocket patient spending.

In response to this bleak picture, there has been increased governmental interest in a fundamental reform of healthcare on a global level. The OECD's June 2010 Report on "Improving Health Sector Efficiency: The Role of Information and Communication Technologies," cites the following key roadblocks to effective healthcare communication as the industry continues to attempt to defragment excessive cost: 1) Purchase and implementation costs for EMRs (Electronic Medical Records) can be significant; 2) Physician incentives differ under different payment systems; 3) Cross-system link-ups remain a serious problem; 4) Lack of commonly defined and consistently implemented standards plagues interoperability; 5) Privacy and security remain a crucial, unsolved problem. The report goes on to say that today's healthcare in the thirty-four OECD countries can see improvement if greater emphasis is placed on quality (instead of volume) of care and on prevention. Furthermore, the report notes that other measures are needed, from reforming cost structures and reorganizing management to equipping and better applying information technology. Finally, one must build "better coherence and knowledge-sharing across often-fragmented health departments, in order to reduce delays, avoid duplication and reduce risks to patients."

In the United States, the latest reform, popularly known as "Obamacare," has been designed to widen access to healthcare for US citizens, but remains a highly controversial political topic and has so far yielded mixed results in practice. Both experience and common sense lead us to conclude that governmental initiatives alone are not the solution.

Taking all of the above factors into consideration, we have designed a comprehensive solution to healthcare management that is far less expensive and involves both the private and the public sectors, on both a local and a planetary level. This solution consists of a fully automated global healthcare exchange, based on our QR technology, which we have called the Zoe Project.

What is Zoe?

Zoe is a massive, QR supercomputer-based, healthcare processing system comprised of software and hardware solutions. It is designed to unify and de-fragment the entire global healthcare process. It is patient-centered, bringing together all healthcare participants (patients, doctors, nurses, alternative healthcare providers, hospitals, clinics, pharmacies, researchers, universities, pharmaceutical companies and insurers) into one framework. All of these entities use innovative web interfaces to access the system from any point in the world. This environment encompasses the personal, public, informational, clinical, and scientific aspects of healthcare. Briefly, the essential features of Zoe include the following:

1. Web-based. Patients, physicians, alternative healthcare providers, hospitals and scientists can access the system from their own computer terminals via the Internet.
2. Tools. Participants can use online decision tools for diagnosis and treatment selection. The system supplies information on preventive healthcare, tailored to each patient. It also provides scientific information and current standards of care.
3. Evidence-based. Treatments are selected based on standards of care, empirical data, and medical literature. Evidence is perpetually updated as new patient results are entered.
4. Central database. The system centralizes medical data storage. Data from individual cases will add to scientific database from which new generalizations and discoveries are made.
5. Advanced QR-based computer technology. Treatment can be monitored with dynamic, interactive flow charts. The system uses advanced QR proprietary data structures that involve meta-data, meta-variables, and meta-database.
6. Communication space. The system organizes communication between patient and provider, and among providers – public or private.

Simply put, Zoe aims to transform the reality of local and global healthcare management by supplying a new, dedicated, empowering web-based infra-structure to support it. The infrastructure can be understood at two levels: At one level, it is a suite of sophisticated applications accessible via the Internet by patients, physicians or healthcare organizations. Thus, it can be understood as a local set of computer programs. At another level, it can be understood globally – as an entire system of such programs. In this case, the total is more than the sum of its parts. The system as a whole provides an unprecedented degree of communication and interaction among patients, providers, organizations, administrators and researchers. The paradigm proposed is the best one to meet the central need of creating an environment in which the adaptive self-organization of the global healthcare system can occur.

Everything we propose with Zoe Systems is feasible using our current QR-technology. There is no question of whether it can be achieved from a techno-logical viewpoint, and eventually, given the proper geopolitical conditions, a system such as this one will undoubtedly be implemented.

Key Healthcare Challenges

In the milestone report, "Crossing the Quality Chasm: A New Health System for the 21st Century" (IOM, 2001), the Institute of Medicine (IOM) outlines the basic principles by which healthcare reform must be approached in the United States. This report represents the opinion of a large panel of health experts on specific actions required to solve the current healthcare crisis. It recommends that private and public purchasers, healthcare organizations, clinicians, and patients should work together to redesign healthcare processes in accordance with the following rules:

1. Care based on continuous healing relationships. Patients should receive care whenever they need it and in many forms, not just face-to-face visits. This rule implies that the healthcare system should be responsive at all times (24 hours a day, every day) and that access to care should be provided over the Internet, by telephone, and by other means in addition to face-to-face visits.
2. Customization based on patient needs and values. The system of care should be designed to meet the most common types of needs, but also have the capability to respond to individual patient choices, preferences and needs.
3. The patient as the source of control. Patients should be given the necessary information and the opportunity to exercise the degree of control they choose over healthcare decisions that affect them. The health system should be able to accommodate differences in patient preferences and encourage shared de-cision-making.
4. Shared knowledge and the free flow of information. Patients should have unfettered access to their own medical information and to clinical knowledge. Clinicians and patients should communicate effectively and share in-formation.
5. Evidence-based decision making. Patients should receive care based on the best available scientific knowledge. Care should not vary illogically from clinician to clinician or from place to place.
6. Safety as a system property. Patients should be safe from injury caused by the care system. Reducing risk and ensuring safety require greater attention to systems that help prevent and mitigate errors.

7. The need for transparency. The healthcare system should make information available to patients and their families that allow them to make informed decisions when selecting a health plan, hospital, or clinical practice, or choosing among alternative treatments. This should include information describing the system's performance on safety, evidence-based practice, and patient satisfaction.
8. Anticipation of needs. The health system should anticipate patient needs, rather than simply reacting to events.

The report also recommended the need for building an information infrastructure that can support healthcare reform. The importance of these recommendations remains unchanged today.

An Appendix to the IOM report is devoted entirely to understanding the problems of modern healthcare from a systems theory standpoint. It notes that the current (U.S.) healthcare system is a complex adaptive system and suggests that one must create conditions in which this system can evolve naturally over time, allowing a wide space for natural creativity to emerge from local actions within the system. The QR-philosophy of redesigning our current healthcare systems is entirely consistent with these statements. It is interesting to note, however, that neither the report nor the Appendix specifies exactly how these recommendations will be implemented. The report supplies a vision, but not a strategy.

We should like to stress, as a basic requirement, that prevention of disease must be a primary target of the new systems. Any comprehensive and self-adaptive healthcare system must provide continuous care instead of "repair medicine," when disease has already erupted or is imminent.

We also fully agree that the new foundation for the system must be based on the most advanced information technology available. It is the only way to monitor the healthcare workflow with the required speed of action and knowledge base. If one accepts this premise, more requirements evolve naturally. All data created in the complete process has to be transferred into the system in an adequate form. This clearly demonstrates why an electronic health record (EHR) is at the very heart of the reform. Currently, government-based initiatives to promote the use of standardized electronic medical records are in progress in many countries. For example in the recent "Plan for a Healthy America," U.S. President Barack Obama gives a succinct summary of the burgeoning healthcare situation, expressing the need for EHRs: "Barack Obama and Joe Biden will invest $10 billion a year, over the next five years, to move the U.S. healthcare system to broad adoption of standards-based electronic health information systems, including electronic health records."

A tremendous amount of work is still needed, however, to make the electronic health record an instrument that supports the clinical workflow, instead of obstructing it. Recently, so-called "medical documentation systems" have mushroomed all over the place. As administrations pump money into rebuilding healthcare systems, everybody wants a share of this funding. Even U.S. supermarket companies such Wal-Mart feel the need to bring such systems on the market. Unfortunately, patchwork solutions of this kind are doing a disservice to the needs of our healthcare systems, routing urgently needed resources into blind alleys. Only an integrated system that is a common platform for all participants of the healthcare process can be a base for success.

Additionally, mandatory digital storage of patient data frees up all possible benefits only if an effective and powerful data-mining process continuously works over the complete database, extracting evidence-based information and alert situations, in real time. For that purpose, one centralized database holding all patient data is crucial. In the era of mobile communication, the integration of the healthcare system into a web-based environment is, by far, the most effective way for easy, continuous and ubiquitous access, guaranteeing the incorporation of the patient as an emancipated partner.

To sum up our discussion so far, the rebuilt healthcare system should have the following requirements in order to be an effective solution for current problems:

1. Generation of a mandatory electronic health record that will constitute a complete and permanently accessible medical history of the patient
2. Instant accessibility for patients, healthcare providers as well as scientists, administrators and insurers
3. Data flow and data storage in compliance with protection of personality and data security
4. Sophisticated "healthcare provider–patient" context-adaptive interface, with up-to-date data acquisition devices to effectively facilitate data transfer to the electronic record
5. Integration of automated patient-monitoring systems with alarm functions
6. Context-sensitive decision tools for diagnosis and treatment, accessible for both healthcare providers and patients
7. Integration of socio-demographic factors such as stressors, geography, family, job and lifestyle data, enhancing possibilities for preventive medicine
8. Instant access to case-related scientific research and standards of care
9. Rapid data back-flow to healthcare providers and patients by continuous hypothesis generation and statistical cross-reference analyzing of the complete database, providing real-time, evidence-based information
10. Easy access to anonymized data by scientific researchers, as well as full integration of medical literature (completing this data with the evidence from the patient record data)

11. Disease cluster observation for rapid detection of new or changed disease patterns, as well as bioterrorism surveillance with an automated, real-time, global monitoring and alert system
12. Common communication space for easy information exchange of patients, healthcare providers and other medical institutions

In what follows we take up again, in more detail, the key challenges facing healthcare today and how the Zoe global exchange system addresses them.

Current Use of Healthcare Information Technology (HIT)

The current healthcare systems involve a disorganized assortment of computer and non-computer systems (paper records). For all practical purposes, they assume a computer-illiterate patient. That is, nothing in our healthcare systems is predicated on the assumption that the patient has a computer and is able to search the Internet for information relevant to their condition. This is clearly a most distorted assumption.

Physicians have computers, usually a desktop computer or local network dedicated to their own clients. Frequently, the only tasks of a physician's computer are billing and administration. Hospitals have mainframe computers, dedicated, again, to billing and administration. Health insurance companies have mainframe computers dedicated to reimbursement issues.

Each individual computer system, in the current arrangement, has its own database. Specific, limited pieces of information are transmitted between computers. But a fundamental problem with this design is that each system keeps an independent "master record" of the patient. Figure 26 below illustrates the current situation for a hypothetical patient:

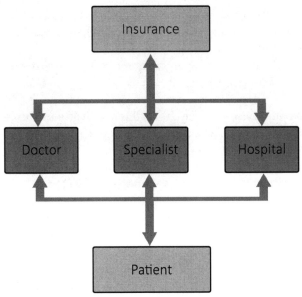

Figure 26. Representation of Current Healthcare System

In this example, suppose a patient consults their primary care physician for a routine health problem. After an examination, the physician refers the patient to a specialist for additional tests. The specialist performs the test, and then refers the patient to a local hospital for a minor surgical treatment. Counting the patient's health insurance company, there are five parties involved: patient; primary physician; specialist; hospital; and insurer (We could easily make the situation more complex by including one or more diagnostic laboratories, regional disease registries, and pharmacies).

Each party, with the exception of the patient, has their own computer record of the patient, all being highly incomplete. The patient supplies redundant information to all parties, and other pieces of information circulate between physicians, hospital, and insurer. However, there is no guarantee over the accuracy or relevance of this exchanged information.

It would be far more efficient to maintain a single, complete, and consistent record of a patient in a central place. That is what the Zoe systems model allows, as shown in Figure 27 below:

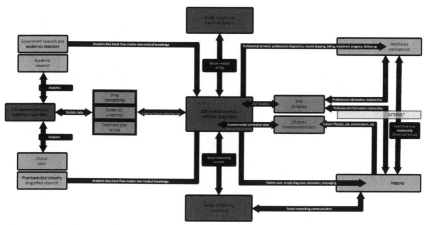

Figure 27. Zoe, the QRT-based Global Healthcare System

Each party accesses the Zoe server, using a web page interface. Using this server, they may supply or view information from the patient's central record – stored on the Zoe servers. They may also run modules on these servers for tasks like diagnostic assistance, treatment planning, drug alerts, etc.

Zoe thus expands the paradigm of SaaS (software as a service), which has distinct benefits in terms of ease of use, no/low up-front cost, and long-term maintenance. (See http://info.softwareadvice.com/rs/softwareadvice/images/ Is_Software-as-a-Service_%28SaaS%29_Right_for_Your_Practice.pdf)

How the Zoe System Works

This is not the place to explain the components and operation of Zoe Systems in detail. Instead, we shall give a general idea of how the system works from the vantage point of a patient, a physician, a healthcare researcher, a healthcare administrator, and a health insurer. As shown in Figure 27a below, access to the system will be granted only after a user authentication, which will be accomplished by modern biometric methods, or username and password identification, depending on the individual situation of the access request and the customer's specific version of the system.

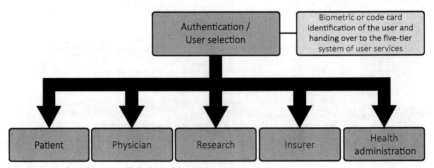

Figure 27a: The Five-Tier Global Healthcare System

After the authentication process a five-tier user group tree will lead to the various functions for the user groups.

The Patient

Figure 27b below represents the Zoe system for the patient's viewpoint:

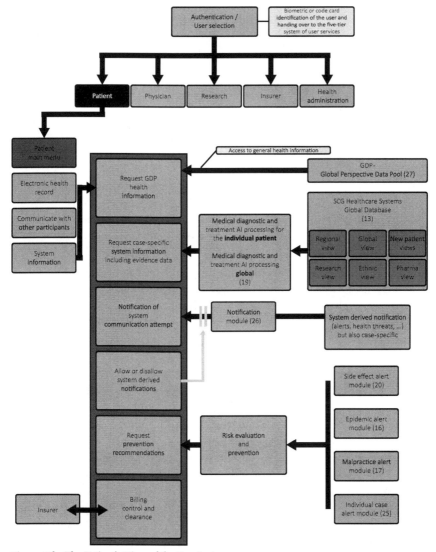

Figure 27b: The Patient's View of the Zoe System

From his or her personal computer, a patient will log on to the Zoe website, using a username and password, or a biometric reader. Once logged in, the patient will be able to perform any of several operations such as:

1. Supply historical information. The patient can supply information on previous medical illnesses, treatments, vaccinations, etc. Information on current health status, life situation, and values and preferences can also be supplied.

2. Organize medical record. The patient can view and organize their medical record. The information in the record is a combination of information supplied by the patient, and records of medical diagnoses and treatments supplied by healthcare providers and diagnostic laboratories. The patient has total control over record security. Any part of the record can be shared or not, at the discretion of the patient. A patient can stipulate that a given physician only be given access to certain aspects of the medical record.

3. Preliminary diagnosis. A patient with a current health concern can enter symptoms into online forms. The forms are customized, asking the minimum set of questions needed to understand the individual patient's present complaints. Based on this information, the computer will attempt to provide a preliminary diagnosis – in terms suited to the patient's level of knowledge.

4. Treatment options. The patient may request and obtain clear information about treatment options. This information will come from established standards of care, as well as from the consumer ratings of treatments retrieved from the Zoe databases and supplied by other system users.

5. Alternatives to medical treatment. Clear, reliable information on alternatives to seeing a physician will be provided. The patient will be given concise and accurate information concerning preventive and alternative health approaches (e. g., holistic, herbal, homeopathic, Chinese traditional medicine, etc.).

6. Select healthcare provider. Should a patient elect to see a physician or another kind of healthcare provider (e. g., acupuncturist, homeopath, naturopath, chiropractor, Ayurvedic medical practitioner, and so on), the Zoe Systems will help them locate one. Physician names, addresses, phone numbers, email addresses, and websites will be provided. A list of potential providers will be sortable by alternative criteria, including geographic proximity, specialization, quality-of-care indices, health insurance compatibility, etc.

7. Appointments. The patient can elect to make an appointment online. If the medical practitioner is a Zoe subscriber, then the patient will be able to view the practitioner's schedule beforehand.

8. Communication space. Once logged on, a patient can choose to access their personal communication space. This area lets providers, hospitals, and health insurance companies communicate with the patient to resolve various questions. In this space, the patient may open a new folder for each new health problem or issue. Within folders, one may open subfolders that apply

to a more specific aspect of a health issue. Folders may be classified as clinical – allowing communication between physicians and patients – or administrative, allowing communication between, say, a patient and their health insurer. Messages between patient and other parties are saved and can be viewed in this communication space. Threads of messages are tracked so that patients can retrieve an entire context of a message.

9. Treatment monitoring. The patient may enter and view information on current and historical medical treatments. For current treatments, the patient may enter information to monitor treatment efficacy and adverse reactions. With the patients' permission, appropriately anonimyzed portions of this information will be made available to their physicians, other patients, epidemiologists, and clinical trials databases.

10. Prevention. A variety of services will be available in the area of preventive health. For example, based on personal genetic, lifestyle, and family history information, the system could assess a patient's risk status for various diseases and suggest a tailored schedule for screening tests.

This range of patient services is entirely realistic and compatible with today's computer technology. In many ways, what we propose here is analogous to an online product such as the Yahoo! portfolio manager (http://finance.yahoo.com/), which allows subscribers to perform a wide array of investment tasks, tracks personal stock portfolios, supplies customized information on stock performance, and links to relevant news stories. It is clearly feasible, if not compellingly important, to have a service with similar breadth of features dedicated to patients' health.

The Physician

Figure 27c below represents the physician's view of the Zoe system:

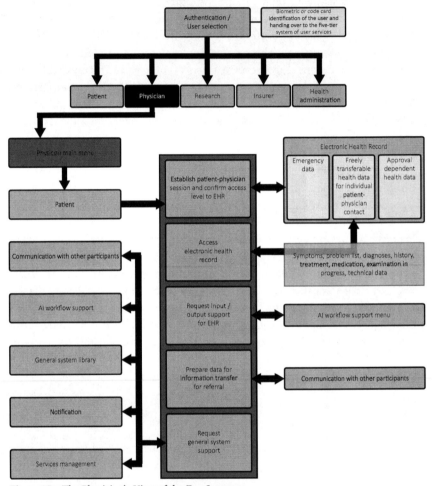

Figure 27c: The Physician's View of the Zoe System

From a personal computer, physicians and other healthcare providers will log on to the Zoe Systems website. They can then perform any of several actions relative to a patient's case, such as:

1. Review patient information. Before an initial visit, they can familiarize themselves with the patient's personal and medical history. They can get a broad idea of the patient's present complaints.

2. Make diagnosis. After examining a patient, the physician may interact with the Zoe diagnostic support tools to help identify the diagnosis. Symptom information is entered using intuitive, adaptive menus that promote efficiency. Diagnoses are suggested, based on the patient's symptoms and medical history, and regional variations in disease prevalence. The latter reflect current epidemiological trends (e.g., presence of a new flu strain).

3. Identify standard of care. Using the Zoe Systems, the physician can locate standards of care relevant to the patient's case.

4. Identify other physicians and healthcare providers to assemble a medical team, if the patient's case requires complex treatment.

5. Make treatment recommendation. The system will make suggestions about plausible treatment options. These options reflect costs, probabilities of successful and adverse outcomes, and individual patient characteristics (e.g. genotype).

6. Monitor treatment. The standards of care, referred to above, include dynamic flow charts. These charts present standards as a decision tree. They are dynamic in the sense that the system can track where the patient currently stands in the treatment process. Knowing where the patient is relative to treatment is an essential technical innovation. It requires advanced, proprietary Enhanced Intelligence (EI) methods that allow the computer to understand the patient and his or her illness. Recommendations for the patient are then issued, based on such understanding. When the patient's treatment deviates from the standard of care, an alert is issued.

7. Incorporate research development. Standards of care should reflect the latest scientific information available.

Figures 27d and 27e below show the physician's QR-based knowledge and interface support, provided by the Zoe system, which keeps the physician abreast of any new developments in his or her field:

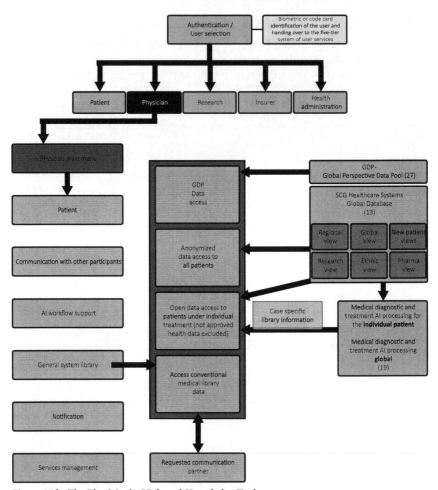

Figure 27d : The Physician's QR-based Knowledge Tools

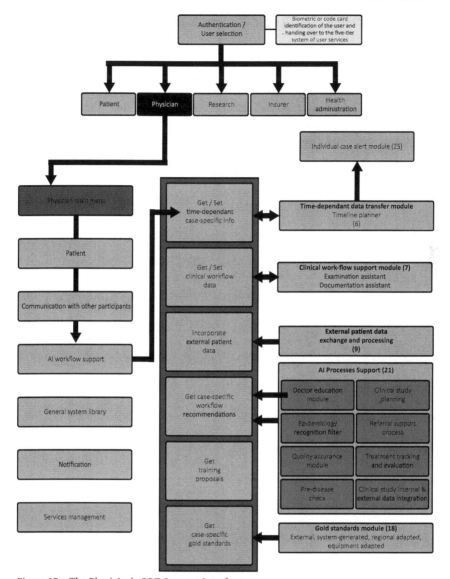

Figure 27e: The Physician's QRT Support Interface

The interface of the system must be a state-of-the-art tool, both comprehensive and relaxing to use. Voice recognition technology is available to allow the practitioner to query all research data relevant to the case at hand.

Zoe as Clinical and Scientific Research Tool

Zoe revolutionizes the way in which the healthcare industry uses information technology to capture, manage, analyze, and exchange patient information. It will enhance patient care, moving medical practice solidly into the realm of genuinely individualized medicine. It functions both as a clinical and a scientific research tool. As a *clinical tool*, the Zoe system addresses questions such as:
1. What is the likely diagnosis of a patient?
2. What are the most effective treatment options?
3. What options seem best for this patient (individualized medicine)?
4. Which treatments are contraindicated?
5. What further diagnostic tests will clarify treatment choice?

As a *scientific research tool*, the Zoe system addresses such questions as:
1. Has the rate of influenza in a given city increased in the last week?
2. What is the probability of success of a certain drug for a patient with a certain gene?

The system gathers scientific information in three ways: First, there is the clinical data supplied by the patients, their physicians, and diagnostic laboratories. Second, information comes from automatically reading, interpreting, and integrating medical journals and related literature. Third, information comes from discovery engines such as Q-search that run on Zoe's central computers. These are sophisticated EI and statistical programs that extract rules and generalizations from the raw data and combine information from different sources (e. g., patient records and published scientific studies). Figure 27 f shows Zoe's research system interface that combines the three methods of gathering and processing scientific information:

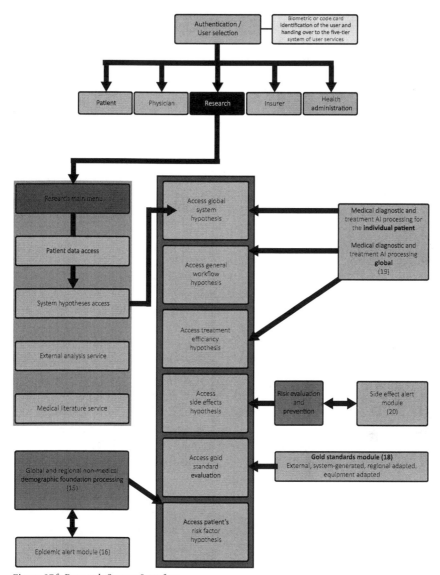

Figure 27 f: Research System Interface

Zoe and Accessibility to Recent Research

Another serious problem of current healthcare systems is the medical knowledge gap, namely, the time lag between a basic scientific discovery and its effect on actual medical practice. An average of about 17 years is required for new knowledge generated by randomized controlled trials to be incorporated into practice, and even then application is highly uneven. (Balas and Boren 2000) Moreover, scientific information of the type that leads to improved medical practice is growing exponentially – far faster than the ability to understand and integrate this information into new therapies and to train physicians in their use. For example, in 1948 there were roughly 4,700 scientific journals. By 1990, the number exceeded 100,000. (Schardt and Mayer 2010) Meanwhile, the Internet has vastly increased the rate of communication of scientific discoveries as well.

Consequently, physicians and other healthcare providers often lack sufficient information about appropriate treatments and standards of care, especially to the extent that these are affected by new scientific results. In turn, the patients do not have sufficient or readily available information about treatments, potential side effects, historical quality of various institutions and physicians (e. g., success and mortality rates), and preventive strategies. Most importantly, "even when good information is available to support healthcare decisions, it often isn't being used to improve care quality because the unaided human mind, no matter how competent, simply cannot focus on all the necessary details nor possess all the knowledge needed for continually making the best clinical decisions. Specialization and traditional information technology do not solve this problem." (Cited at http://wellness.wikispaces.com/The+Knowledge+Gap) All of these factors produce an ever-widening medical knowledge gap.

Another serious, related problem is the ever-increasing costs involved in clinical trials. In the United States, one new pharmaceutical drug candidate costs an average of $800 million dollars to bring to the market place. With the rapidly rising potential of new drugs in the development pipeline, we can no longer afford this massive price tag attached to the admission and authorization of new candidates. There is no denying that the randomized controlled clinical trial – the mainstay of the current system for evaluating new drugs – has been very valuable. But the times are changing, and so must our methods.

A new paradigm for clinical trials is mandatory. Observational clinical trials that evaluate new drugs should be seamlessly interwoven with clinical practice, thus playing a greater role in the process of drug evaluation methods. Instead of a clinical trial performed at a few selected sites, we need trials in which, in principle, any physician and any patient can participate, regardless of their geographical location. The centralization of records, data, training, procedures and outcomes will, in turn, be handled electronically.

The danger of new medications is that adverse and potentially lethal side effects are often revealed only after a prolonged market presence. Obviously, this creates a time lag whereby the harmful medicine has already been deleteriously administered to thousands of patients. A paradigm shift in clinical research has two prerequisites:

1. The information culture starting at a university hospital and ending in the most basic medical facility has to have a quality that makes it possible to sum up every bit of medical data for a joint observational study.
2. Sophisticated and responsive alert systems that continuously monitor the entirety of all treatments and analyze the complete data pool for unknown adverse effects of medical processes are required.

As a prerequisite for an effective implementation of these demands, we need a new methodology with centralized and rapid access to all patients' data. An anonymization of this data is mandatory to protect the patients' privacy. The Zoe system is designed to meet all of these needs. Zoe converts the overflow of the world's medical information into relevant and accessible knowledge. A massive medical database will include knowledge elements of all relevant types (i.e. genomic, environmental, demographic, etc.) The system uses sophisticated EI methods for representing and applying common sense rules about how variables interact in the real world. It acts as an information filter for critical situations, delivering vital information to individual physicians and other healthcare practitioners.

Furthermore, Zoe allows pharmaceutical researchers to identify the patients throughout the world who present just the characteristics needed for their clinical trials. The system can coordinate clinical trials by bringing together willing patients and researchers, dramatically improving both the efficiency and the scientific relevance of clinical trials. Thus, because of Zoe's ability to accumulate vast amounts of medical data, clinical trials will soon become available to virtually anyone as a result of dramatically reduced cost. It should be noted just how much this value alone can begin to create a paradigm shift in healthcare.

The Health Insurer

Figure 27g represents the health insurer's view of the Zoe system:

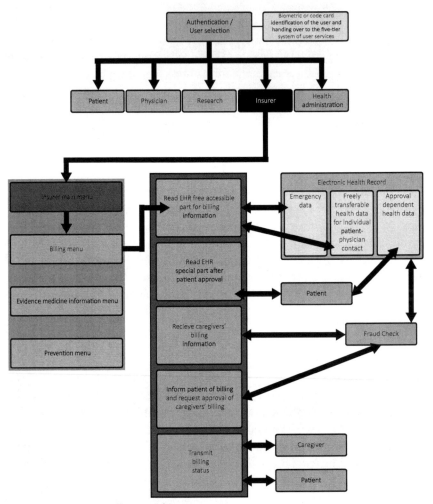

Figure 27g: Insurance Billing Interface

Computers from health insurers can connect with the Zoe server electronically. Reimbursement claims for procedures that eligible patients receive will be supplied automatically. Data relevant to the claim (ICD 10 diagnosis, procedure code, visit date, etc.) will be supplied. No data will be transmitted to the insurer other than that which is authorized by the patient and/or physician/healthcare provider and is directly relevant to the current claim. Case workers may manually log on to the Zoe site. There, they may review authorized portions of the patient's history. This will assist them in providing customized, personal contact, in cases where that is deemed warranted. (Many insurers currently provide such contact for certain chronic diseases, having determined that by promoting timely medical interventions and preventive measures, this is both cost-effective for the insurer and beneficial for the patient.)

Zoe will enable insurers to tabulate summary statistics on regional patterns of care. Irregularities by individual physicians or other healthcare providers (indicative of fraud) can be detected automatically.

Zoe and Health Administration

Figure 27h below shows the Zoe system from the viewpoint of the Health Administration:

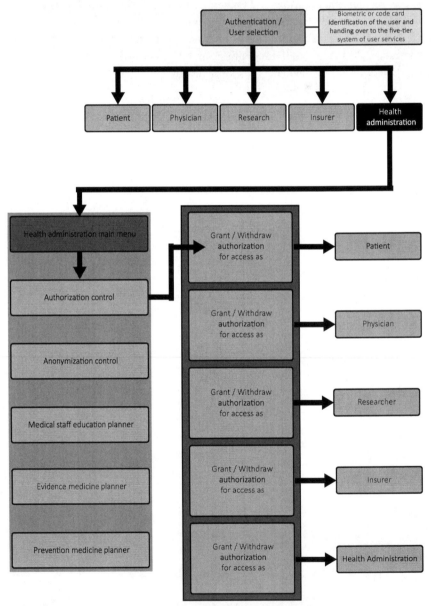

Figure 27h: Strict Authorization Control for Patient Privacy Protection

Health Administrations must ensure the strictest standards of patient privacy and the highest standards of medical research and healthcare available, including preventive medicine and food and drug policies. The Zoe system provides unique support in maintaining and/enforcing such standards, as we shall describe in the subsections on patient privacy and preventive medicine below.

The Problem of the Healthcare Record

Modern medicine would be impossible without patient healthcare records. A medical records system provides, at a minimum, accurate and systematic record keeping of patients' age, occupation, diseases, as well as the modes of diagnosis, treatment history, and recommendations. Each interaction between a patient and a healthcare provider produces valuable medical information. But, today's healthcare practice, even in the most advanced of countries, is mostly kept on 'dead' paper records or disconnected local computer databases, where it stays silent until looked at when a patient is being examined, and is never evaluated against a meaningful pool of comparative data.

When Benjamin Franklin invented modern medical record keeping in 1752, his system was adopted throughout the Americas and then throughout the modern world. Between 1752 and 1967, medical record-keeping technology remained essentially unchanged. Indeed, today thousands of hospitals have rooms full of paper records that would be instantly recognizable to Franklin as the stepchildren of his first system. In 1967, Dr. Lawrence Weed and the University of Vermont created the first electronic record-keeping system (PROMIS). The developers of PROMIS quickly realized that keeping records electronically meant that they could also be analyzed electronically, so that in 1970 they developed the POMR system to add diagnostic and treatment plans.

The early PROMIS (Patient-Reported Outcomes Measurement System) and POMR systems had as explicit goals to use the medical records not only for patient treatment, but also for the rapid collection of data for epidemiological studies, medical audits and business audits. Today, medical records are routinely used for business audits and less frequently used for medical audits. The goal of data collection for epidemiological studies has largely been ignored.

HIMSS defines EHR as follows: "The Electronic Health Record (EHR) is a longitudinal electronic record of patient health information generated by one or more encounters in any care delivery setting. Included in this information are patient demographics, progress notes, problems, medications, vital signs, past medical history, immunizations, laboratory data and radiology reports. The EHR automates and streamlines the clinician's workflow. The EHR has the ability to generate a complete record of a clinical patient encounter – as well as supporting other care-related activities directly or indirectly via interface – including evidence-based decision support, quality management, and outcomes reporting." (www.himss.org/ASP/topics_ehr.asp)

Electronic health records (EHR's), like all computer data, are useless unless they are structured and organized. Precise coding must be used, and that coding must be standardized across patients and across time. In order for different healthcare providers to use these records, there must be consistent and universally recognized standards. And yet, it is a common saying in the medical world that "The nice thing about medical record keeping standards is that there are so many from which to choose."

Traditional EHR (Electronic Health Record) systems are both capital intensive and expensive to maintain. The typical cost for a fully featured EHR system is about US $35,000 per physician, so that only the practitioners in the G20 can generally afford to acquire them. This puts such systems out of the reach of most practitioners in other countries. At the same time, there are thousands of pages of regulations in the United States (and similarly in other G20 countries) regarding EHRs, as well as tens of thousands of partially implemented, out-of-date or obsolete EHR systems, contributing greatly to healthcare costs, besides rendering them inefficient.

Zoe and EHR's

Zoe is built using the most modern data processing technologies, including the latest EHR standards. Although it interfaces with legacy records systems, it is not constrained by an obsolete design. Zoe is to traditional electronic records management what Google, Facebook, and Wikipedia are to a dictionary or thesaurus: it provides many of the same services (basic medical record-keeping and billing support) plus data collection for analysis as well as personalized patient pages that offer lifestyle support and social networking functions. Because it is designed for global, not just local, implementation, Zoe could become the ultimate Gold Standard for healthcare actions and treatment plans throughout the planet.

Zoe's Electronic Health Record (EHR) is initially created by the patient, or by a healthcare provider on behalf of the patient. The creation of this record begins the 'enrollment process'. Ideally for security, the patient uses a high-end, Zoe-provided, USB fingerprint (biometric) reader to secure access to the information inside his or her own EHR. The enrollment process is free and open to anyone who chooses to participate. Each patient and his or her authorized healthcare provider can add healthcare information and measuring tools, from thermometers to CT Scan equipment. In the most ideal case, the EHR will be a true representation of the entire healthcare history of each enrolled person.

The Zoe supercomputer and the patient's record make a very effective clinical diagnostic support possible. This support system, which we call the Zoe Diagnostic Support (ZDS), analyzes all of the facts inside each EHR and offers diagnostic or treatment suggestions to the healthcare provider(s). The ZDS follows each of the treatment results in each EHR, as entered by the healthcare professionals or augmented by the patient himself and, over time, learns the positive and negative results of its treatment programs. This computer knowledge is unique, because it is derived in real time from many millions (and eventually billions) of records of patient data. The supercomputer performs continuous statistical evaluations of all recorded treatments, checks for drug interactions, and searches for any other correlating factors that can be found within the total pool of EHR records.

One should add that there is a ready market for EHRs: "The number of national electronic health record initiatives around the world is certainly growing in leaps and bounds. Today, there are at least 23 countries where national electronic health record strategies are in place for the entire country and these strategies have been or are being implemented. In alphabetical order, the 23 countries include: Australia, Austria, Belize, Canada, Croatia, Denmark, Estonia, Finland, France, Hong Kong, Japan, New Zealand, Netherlands, Norway, Saudi Arabia, Singapore, Spain, Sweden, Taiwan, Turkey, U.A.E., U.K. and the U.S.A. Of these 23 countries Denmark is widely recognized as the global leader." (Cited in http://infowayconnects.infoway-inforoute.ca/blog/electronic-health-records/170-national-electronic-health-record-initiatives-%e2%80%93-the-2011-who%e2%80%99s-who/#ixzz1JL0qCee9) Zoe's system will not compete with, but seamlessly integrate these local EHR's, using its Quantum Relations concepts of DFO's and frames of reference in order to build its global database.

Zoe and Patient Privacy

The EHR of Zoe is unique, because it allows in its databases not only standardized medical information, but also data relating to ethnicity, diet, religious customs, lifestyle choices, environmental and geographic conditions, education and other secondary factors that could influence the health of a person. Rigorously enforced privacy standards keep this information within the patient's control, releasing it only with his/her authenticated permission.

Patient privacy is paramount, and the ZDS operates only on abstracted patient data without identifying personal information. Therefore, the derived knowledge can be made instantly available to healthcare professionals, as well as to researchers, academia and to the pharmaceutical industry on a subscription basis. Each EHR is analyzed in real time, regardless of whether new information has been entered or not, because the system automatically applies new knowledge to the analytical process and therefore can update the diagnosis of a patient wherever and whenever new and case-relevant knowledge is created. This was, in fact, a primary goal of early EHR systems, but it has largely been forgotten as the EHR industry developed fractured record standards and became increasingly dominated by efficiency experts and insurance companies.

Zoe, Preventive Medicine and Health Literacy

A basic problem in today's healthcare system is the marginalization of the patient. Patients are often passive spectators, rather than active participants in the process of disease management. It is clear that the current disjunction between the patient and the process cannot be sustained over the long term. The crushing cost of healthcare mandates that preventive health, a model best organized by the patients themselves, play a greater role. Similarly, individual patient preferences and values are insufficiently considered in current healthcare practice.

The genetic revolution will emphasize the need for the individualization of medicine. The shift to a personalized medicine will most certainly lead to more specific diagnoses, individualized treatments, and new considerations regarding a patient's unique genetic background, which will be uncovered by new genetic testing techniques. Additionally, socio-demographic, cultural, job, family and other patient data have a crucial effect on the outcome of care giving. These variables are currently not adequately included in the healthcare process. Only by strengthening the role of the patient as an active participant in the system, as well as by the creation of a responsible partnership between patient and caregiver, is an ideological-shift towards continuous health preservation (as opposed to disease-repair) made possible.

The Internet now gives many patients access to a great deal of information regarding medical facts. Patient advocacy groups maintain detailed, informative websites for nearly every major disease, but frequently lack a personalized attitude. Anyone can review the current medical abstracts (e.g., via the U.S. National Library of Medicine's PUBMED service). The healthcare delivery system must accommodate and exploit the open source availability of medical knowledge. Figure 27i shows Zoe's sophisticated preventive healthcare interface system:

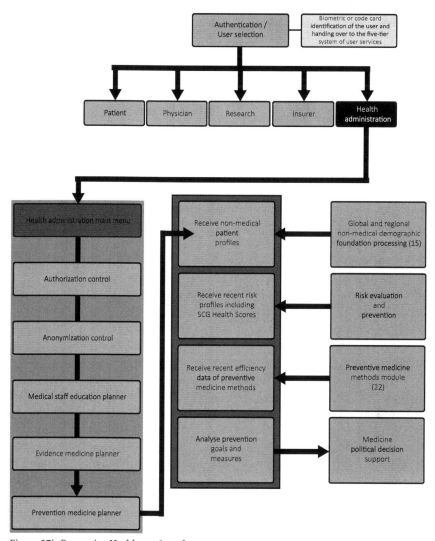

Figure 27i: Preventive Healthcare Interface

Preventive medicine presupposes a high degree of health "literacy," which is perhaps one of the greatest contributing factors to ineffective healthcare throughout the world. Health literacy can be defined as an individual's ability to read, understand and use healthcare information in order to make decisions and follow instructions for treatment. For example, US studies reveal that "up to half of patients cannot understand basic healthcare information." These studies note that "low health literacy reduces the success of treatment and increases the risk of medical error." In turn, "various interventions, such as simplified information and illustrations, avoiding jargon, 'teach back' methods and encouraging patients questions, have improved health behaviors in persons with low health literacy." The studies conclude that health literacy "is of continued and increasing concern for health professionals, as it is a primary factor behind health disparities. The government's Healthy People 2020 has included it as a pressing new topic, with objectives for addressing it in the decade to come." (Cited at http://www.healthypeople.gov/2020/topicsobjectives2020/overview.aspx?topicid =18)

In addition to the adverse effects of low health literacy on the individual patient, there are economic consequences to society as a whole. The National Academy on an Aging Society estimated that, in the United States, additional healthcare costs due to low health literacy were about $73 billion in 1998 (Cited at http://www.agingsociety.org/agingsociety/publications/fact/fact_low.html).

It is our belief that if the improvement of health literacy were left solely to the global public health infrastructure, as it exists now, the evolution of people's health knowledge and subsequent 'literacy' would be a very slow process. Many industry experts maintain, and the statistics support them, that poor health literacy is a result of poor distribution of knowledge. It is largely left to governments to produce curricula and materials that are most often distributed only through patient visits with care providers, who then relay some form of knowledge to the patient. It is therefore clear to see why health literacy is so detrimentally poor and why it has such a dramatic cost associated with its failures.

With Zoe, the patient can monitor all aspects of healthcare and will quickly become health literate. For this purpose the system develops the Zoe Life Partner, available online to its patient-users as an entirely interactive and entertaining health and healthcare ecosystem. Patient responsibility, especially in the area of preventive health, is emphasized and personal preferences are respected.

With an emphasis on educating the patients of the world to better understand all of the virtues of health and healthcare, Zoe will move the healthcare process from reactive to proactive, assisting physicians to implement state-of-the-art preventive care. The biggest complaint of doctors today about EHR systems is that they inhibit and restrict doctor-patient communication. Zoe does exactly the opposite, helping the physician by giving a simple communication channel over the Internet. This approach is almost unknown to the medical community today, but is well known to every user of Facebook or any other social networking systems.

Thus, Zoe is the first complete health ecosystem that creates an entirely self-contained global community. Zoe provides a personalized space (much like Facebook, LinkedIn, MySpace), completely customizable by the patient/person.

It connects the patient with a global database of health and healthcare knowledge and actually becomes a virtual lifetime 'friend' as it monitors the member's medical life and other significant, health-related factors, such as daily vital statistics.

Other Advanced Features of the Zoe System

Figure 28 below shows an overview of Zoe's entire QR-based technological platform:

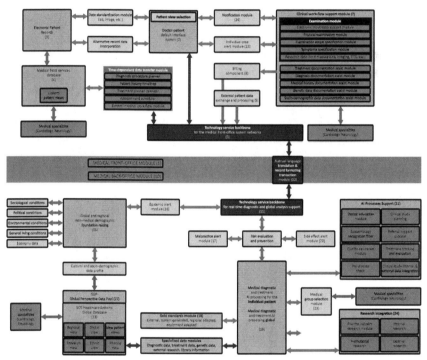

Figure 28. Overview of ZOE System

In addition to the features we have already described, the Zoe system has a few other important advantages such as:

Compatibility and Interoperability

A serious systemic problem with HIT today is that there are literally thousands of different standards by which the various healthcare providers communicate.

This reduces the providers' ability to share critical data and, therefore, to provide the best care possible (misdiagnosing/mistreating as a result of lack of knowledge/information of a patient's medical history). Zoe is the first system capable of full compatibility with any of the practitioner's existing system(s). It provides for interoperability amongst healthcare givers globally, allowing for critical information to be shared, in real-time, amongst all participants enrolled in the system.

Interactivity

The Zoe system creates a virtual 'living' environment whereby any user can log-in to the system and find immediate value, as well as become engaged with the system in a pre-defined set of ways. For example, Zoe is designed to take a user's blood pressure, temperature, weight, and so on at a preset time every day, with automated alerts generated to notify the user of dangers, statistical data, nutritional advice, recipes, potential problems that should be addressed with their practitioner, and so on. Via automated wireless devices, Zoe is even capable of alerting emergency responders to a patient who has, for example, stopped breathing in a given environment (home or wherever an apparatus is in-place), should that patient's profile provide this level of directive.

Integration

Zoe can also act as a clearing-house of sorts in that it offers comprehensive integration features. Healthcare providers need not replace any of their existing hardware or software solutions, but rather can seamlessly 'plug-in' to Zoe and use whatever features they choose.

Billing

The patient's records are easily moved from one doctor to another, because they are stored centrally (where local practice permits). Doctors are more assured of payment because billing is no longer only the problem of one or two medical technicians but is part of an integrated system. The EMR system doctors currently use can be easily integrated into the Zoe system to communicate with payers. The payer groups (insurance companies or governments, for example) will receive standardized bills that do not have to be manually re-entered into an EHR system.

Social Networking

As we have already mentioned, the system has a social networking element that allows Zoe users to connect with each other, on the basis of any chosen set of common interests, whether they are persons with a common affliction seeking a support group or wanting to share a common experience, or someone looking for exercise tips. This is a proprietary feature.

Jobs

Zoe's function will include a global healthcare jobs database, free to both employers and employees wishing to communicate their respective interests.

Costs

There is no cost to Zoe's patient members, and the system requires zero capital investment on behalf of the practitioner (by contrast, in the United States, the most sophisticated, full-featured EHR software options – nearest to Zoe in feature dynamic – have an average associated purchase price of $85,000 per practitioner). Zoe will, of course, charge the medical providers for transmitting or retrieving records, but these costs are pennies per patient per month. Zoe will make its profit from the volume of such transactions, which will eventually be in the millions per day.

Zoe and Cost Savings in Local and Global Healthcare Markets

Zoe can act as a comprehensive communication sphere for medical providers, medical and healthcare practitioners, hospitals, clinics, research institutions, and patients. It supplies the medium for communication among all parties, including transmission of x-rays, results, files, etc. It can also broker consultations and second opinions, all communication being electronically documented. In addition, Zoe provides the communication backbone needed for a global response to epidemics and other variables adversely affecting health over large geographical spaces.

The Zoe approach to local and global healthcare will thus provide significant savings of total healthcare costs in the countries where it is adopted. Estimating the savings is difficult for many reasons, including vastly different levels of existing IT infrastructure, vastly different healthcare delivery realities (e.g., there is one doctor per 7,700 patients in Indonesia, and one doctor per 320 patients in the Netherlands), and vastly different resources available to implement any new technologies or medical methodologies. However, logic must allow that the raw reduction of inefficiencies across the board, the availability of real-time data, the reduced incentive to engage in fraud and the education at every level of membership, will account for savings within the 'system' in every market where adapted. The greater the level of adaption, the greater the savings as economies-of-scale begin to take effect.

The Zoe approach takes into account these factors by providing a modular set of solutions. First, both the patient interface and the doctor's interface are implemented using web technology that will run on any available computer system, including legacy computers, hand-held devices, and even the ultra-low cost OLPC XO laptop system, which is available to students in developing countries at a cost of less than $100.

Second, although the system internally uses a specific model of healthcare records (presently openEHR with extensions), interface and translation modules will convert the internal representations of healthcare record documents into every commonly used standard, including clinical standards such as HL7, ANSI X12, CEN, CEN-EHRom, and so on.

We have chosen openEHR as an internal standard for healthcare records data representation primarily because it separates information services (e.g. data aggregating) and enterprise viewpoints, giving a service enterprise approach particularly suitable for Zoe, and because it permits local variations in practice to be incorporated in an orderly fashion, much more so than the U.S. developed standard, HL7.3, for example.

Within this framework, we can implement patient interfaces and doctor interfaces that are suitable to local conditions. We think globally by having a centralized system for performing difficult tasks of data analysis, data security, and data communication, but act locally by implementing the user interface modules differently for each country, and within each country, for each identifiable group of users.

In the Netherlands, for example, HL7v3 is a nationally adopted standard. There is a specific Dutch infrastructure (AORTA) and an established system of unique patient identifiers. The Dutch government has historically chosen to pursue distributed data storage, but to permit access to the local data through a central registry (the National Hub). To implement Zoe and still comply with Dutch requirements and preferences, we can easily convert back and forth between openEHR and HL7v3, but we must provide local storage for data and a transparent interface with AORTA. We do not wish to compete with the existing Dutch EHR system, but rather to complement it. That means that in this case savings will come from the non-traditional functions of Zoe, including reduction of medical errors; drastic reduction of billing fraud; low-cost access to medical data and research; interactive patient education; and enhanced direct patient-physician communication.

Quantification of these benefits in terms of percentages of healthcare costs is difficult and subjective, particularly where EHR implementation is well under-way and the systems are mature (as in the Netherlands). Nevertheless, based on comparable studies in the United States and elsewhere in Europe, the in-troduction and use of the Zoe system in one of the G20 countries (such as the Netherlands) is likely to reduce overall medical costs by 15–20 %.

At the opposite end of the healthcare spectrum, for example, in Indonesia, Zoe's benefits will be even more tangible and much larger. Most doctors do not have any sort of electronic medical records, although most doctors have, or could afford, a low cost personal computer. Because the Zoe system requires zero capital investment by the doctors (while other systems' computer servers cost upwards of US$20,000 while vendor-supplied, end-user terminals cost $2,000 each), it is affordable. As previously mentioned, a 20-year old PC with a serial biometric interface will support the core functionality of the provider's patient and records interfaces. Faster computers, better displays, and higher speed connections are useful for some of the more exotic functions of Zoe (medical imaging, for example, or transmission of CT Scan data); however, a strong emphasis will be placed on simplicity and availability.

In Indonesia, there is one doctor per 7,700 patients. With this situation, implementing EHR is critical to providing any meaningful basic medical care. Keeping accurate paper records for 7,700 patients is impossible, so that patients either receive no medical care at all, or receive care without any meaningful history. We do not claim to be able to solve the problem of seventy thousand doctors delivering medical care to 240 million people, but we can provide low cost, standardized support for those seventy thousand at a reasonable cost. Sixty-four percent of the total doctors are located in Java, where Internet services are readily available. The installation cost of the system will be minimal – many doctors already own computers, and a web-based solution means that the doctor does not need to hire an IT specialist, a database specialist, a coding specialist, and a network specialist just to use our EHR system.

According to a 2005 benchmark survey of 3,300 practices conducted by the Medical Group Management Association in collaboration with the University of Minnesota and funded by the U.S. Department of Health and Human Services, in the United States the average cost of implementing EHR (hardware and software) is over US$35,000 per physician, while the average annual maintenance cost is over $800. These figures do not include staff training time, and they do not include the costs of making electronic claims for payment. For Indonesia, such figures are impossible for doctors who on average make less than $1,000 per month. For a Zoe system, the cost of implementation is zero, and the maintenance cost of the system is essentially zero. There are, certainly, costs to the medical providers for transmitting or retrieving records, but these costs are pennies per patient per month.

For our Indonesian doctor, the savings from a Zoe system are clear. There is an upfront savings of $35,000 and a monthly saving (for comparable services) of $65, offset slightly by a medical system (government or insurance) paid per-patient charge of (typically) $0.20/patient x 22 (patient days per month), which is less than $96/month, for a first-year savings of more than $22,000.

The Zoe System: Major Benefits

In this section, we briefly recapitulate the ways in which the Zoe system will directly and effectively address the problems with current healthcare:

1. *Cost:* Zoe reduces healthcare costs by making healthcare more efficient; reducing medical errors, reducing physician time burden, promoting prevention, eliminating redundancy, reducing administrative costs, and lowering the cost of clinical trials.
2. *Fraud:* Zoe virtually eliminates fraud by keeping a record of treatments and interactions, and by providing the requisite visibility to all parties involved – patient, insurer and practitioner.
3. *Quality of Care:* Zoe will improve quality of care by promoting and improving evidence-based medicine, the adherence to Standards of Care, and inclusion of patient values and preferences. Error monitor sends alerts when mistakes are evident.
4. *Inequity:* All patients have continuous access to the information of the highest quality. Patients can compare treatment they receive to the Standards of Care.
5. *Physician shortage:* Zoe will promote a more efficient use of physician clinical time. Automated decision tools improve physician productivity and greatly reduce administrative burden.
6. *Medical records:* Zoe eliminates redundancy. Once a patient fills out personal information, it is automatically stored and interpreted permanently. This is extremely helpful when a patient visits a different hospital, clinic, ward, doctor, or specialist. At the same time, a patient reserves the privacy of revealing only the information that he deems relevant.
7. *'Communication:* Zoe acts as a comprehensive communication sphere for medical providers, medical practitioners, hospitals, clinics, research institutions, and patients. The system supplies medium for communication among all parties, including transmittal of x-rays, results, files, etc. Zoe

brokers consultations, second opinions, and assembly of medical teams. All communication is electronically documented.

8. *Knowledge gap:* Zoe will convert the overflow of the world's medical information into relevant and accessible knowledge. A massive medical database will include knowledge elements of all relevant types (i.e. genomic, environmental, demographic, etc.) The system will use sophisticated EI methods for representing and applying common sense rules about how variables interact in the real world.

9. *Global database:* The accumulation of medical data in innumerable patient interactions throughout the world is of immense value to humanity. Zoe's arguably greatest long-term achievement will be to tap into this reservoir of data for the greater benefit of individual patients and scientific research.

10. *Patient involvement:* the patient controls his or her medical record and can monitor all aspects of healthcare. The system emphasizes patient responsibility, especially in the area of preventive health, and respects personal preferences and choices.

11. *Inclusive and integrative philosophy:* Zoe's medical philosophy is based on the integration and harmonization of the various concepts and practices of medicine. The system supports complementarities of, rather than the competition among, various successful healthcare practices from around the world. Thus, it facilitates the cooperation (and eliminates the conflict) between Western and other healthcare systems, while promoting the highest standards of healthcare, as required by evidence-based medicine.

Continuing Development of the System

Once the basic model of Zoe is implemented, every personal computing device (including desktop computers, laptops, tablets, and PDAs) becomes a potential Zoe terminal. The Zoe project includes the design, implementation and distribution of an ultra-low-cost interface to digital medical equipment. The interface will initially be wireless and USB/mini-USB, but we anticipate better interfaces as they emerge on the market.

A basic wireless interface will be a networking device consisting of one or more receivers and one or more transmitters. We anticipate that these will be based on RF MEMS and low power CMOS. They will be small enough to fit easily inside a thermometer, and the same biometric security device that reads the Zoe user's biometric identification will contain a wideband programmable receiver, capable of communicating with these devices.

Applications in long-term patient monitoring are obvious. Less immediately obvious are the advantages of obtaining, from the user himself or herself, accurate readings of temperature, pulse, blood pressure, and respiration. At each point where a human being, even a highly trained nurse, intervenes to transcribe a reading from the medical device to patient records, there is the possibility of error. The elimination of these errors nearly justifies, by itself, the costs of such interfaces.

Adding the RF interface to each such device will cost pennies. In addition to in-home medical monitoring devices, countless medical devices at the clinic or the physician's office can be directly interfaced to a Zoe system, including, for example, patient emergency alarms, thermometers, blood pressure measurement/ monitoring, blood glucose monitors, pulse oximeter devices, pedometers, cameras, medication dispensing devices, in-home EKG devices (cardiac chest strap), genome mapping device(s), personalized prescription calculator, stethoscopes, scale/height measure, Zoe MedPad, and mobile applications ('apps').

The Zoe medical provider terminal devices will contain the biometric identification module, plus an interface capable of supporting Bluetooth, Zigbee, 802.11 wirelesses and other required protocols, as well as a USB (and eventually Thunderbolt) interface.

From the start, Zoe will deliver personalized wellness summary pages to each registered healthcare recipient. This could be as simple as a daily weight chart with an integrated calendar for doctor's appointments or could add prescription drug information, or a complex social networking support group for patients with similar conditions (diabetes, for example). The idea is to make healthcare less fragmented, more integrated for the patients, and to encourage the patients to use the Zoe system regularly in their daily lives.

At the same time, the data gathered by Zoe will be used to increase medical knowledge. The standard method for testing new treatments is the clinical trial. This method can be supplemented and, in many cases, actually replaced with computer analysis of large-scale databases of clinical observations. There are obvious problems with this approach, including bias in treatment assignments, non-standard definitions, improper specification of control and test groups, missing or corrupted data, and statistically invalid comparison methods (e.g., multiple use of the same data), but Zoe is designed to minimize these problems and, in all cases, to make the problems more visible. To give one example, internally Zoe uses the OpenHCR coding for medical diagnosis, but where that coding is derived from translating another healthcare standard, this fact is flagged to the researchers.

Through the use of strict standards and because the implementation team includes physicians, researchers, and healthcare professionals, the methodological problems of using large databases for testing new medical treatments can be overcome. The problems are essentially statistical and are amenable to statistical methods of correction.

Implementing Zoe locally and globally

Whether this kind of healthcare technological platform will be successful or not – and here, we do not mean primarily technological or financial success, but, rather, success in furthering human development – will depend, again, on the quality and nature of its databases and the principles that inform their collection and processing. A project of this kind could go either of two ways: it could content itself with creating a healthcare and medical research technological platform based solely on the principles and practices of mainstream, Western, "allopathic" medicine; or attempt to integrate the perspectives and methods of all other traditional and nontraditional forms of medicine, such as Chinese, Ayurvedic, homeopathic, and so on, as we have proposed in this chapter. In the latter case, the project would become considerably more complex (and more expensive), but also much more worthwhile, having the potential of entirely remapping global medical knowledge and healthcare practices.

In either case, however, one would have to take into consideration local medical research and cultural practices, which differ vastly even within the reference frame of what to some might appear as a unified body of medical knowledge, whether allopathic or not. For example, simple medical facts such as a patient's temperature, diet, blood pressure, pulse taking, blood work, etc. are attributed different values in different allopathic subcultures. Other nonmedical, cultural problems will involve synchronization of different methods of collecting, storing, and interpreting data, synchronization of national and/or local administrative and legal systems related to healthcare, and so forth.

The proper resolution of such issues will necessitate intercultural and crossdisciplinary teams, composed not only of healthcare providers and administrators, medical and biological researchers, and computer scientists, but also legal researchers, cultural anthropologists, historians of science, cognitive psychologists, linguists, sociologists, ecologists, politicians, conflict resolution specialists and other negotiators. But, the steepest obstacles that a project of this kind will face will certainly not be of a technical, or even of a cultural nature. Rather, they will concern a certain human mentality that is far from being restricted to the West. We are fully aware that EHRs and medical services in general are not only highly fragmented but also highly politicized with entrenched constituencies, because healthcare services represent tens of trillions of US dollars.

Therefore, the project will, above all, involve skillful negotiation and harmonization among various powerful special interests and will need to address, in a creative way, the current, worldwide exemptionalist, protectionist, and territorial practices, not least those of the big multinational pharmaceutical companies, without whose cooperation this project has very little chance of success. One will thus need to target specific countries and specific populations, and to know how to operate in those markets. One can also market the system directly to the patients – an approach that has been dramatically effective in the pharmaceutical industry and should be in this case as well.

At any rate, there is no question that Zoe can be achieved from a technological viewpoint today. Given the proper geopolitical conditions, we have no doubt that Zoe, or a system like it, will be fully implemented at some point in a not too distant future.

Chapter 8.
Planetary Network of Intercultural Centers for Integrative Knowledge, Technology, and Human Development: Managing Global Learning and Research in the 21st Century

In the previous two chapters, we have described the electronic architecture of two projects, based on the Quantum Relations Principle, which if fully implemented could lead to radical changes in the domains of local and global commerce and healthcare. But we are not there yet, for various reasons, not the least one being the poor utilization, on a global level, of the technology that we already have in place. Despite the fact that our current information and communication technologies could, as we have seen in the present book, play a decisive role in generating, remapping, and integrating knowledge across disciplinary and cultural boundaries, they are hardly integrated themselves. For the time being, they remain, to a large extent, mere appendices and tools haphazardly filling immediate, pragmatic and conflicting needs, without adequate concern or debate about their long-term development and effects.

One of the greatest weaknesses in the current usage of the ICT global network is that each entity collects data separately (in both the private and the public sectors), without a common data management and structural standard, with incomplete and incompatible analytical approaches, and very little contextual data structures. In the case of governmental systems in the USA, for example, the digital global network is hampered or even blocked by interagency secrecy, jealousy and conflict, as well as by the peculiar funding strategies of the US Congress, while in the case of the American private sector it is blocked by unrelenting competition and blind greed. In both cases, the result is a system of fragmented components that, although presently unrivaled in the world, serves only the interest of a handful of officials and investors.

We are not sure, therefore, that we have quite as yet reached a juncture, in the collective life of humanity, at which we are ready to undertake the radical worldwide reforms needed to unleash our true human potentialities, even though many of these reforms and their promise are encoded in the collective wisdom of our world civilizations and even though we can relatively easily and inexpensively built the technological platforms to support them. We do believe, however, that the next two or three generations can at least prepare the way for such reforms, that is, prepare the conditions that will allow their worldwide emergence and implementation. It is for this reason that humanity's common project for the next few decades should be an educational one. And its main objective should be the creation of local-global learning environments throughout the world, which will open us toward our true human possibilities. It is high time for the open-source ICT community, together with the academic and other epistemic communities around the world to pool together their resources and start moving in the direction of a holistic, digital global network of knowledge, facilitating, instead of hampering, its full emergence.

One must begin to create a globally continuous and multilayered, fabric of knowledge-enabled networks, application hosts and open-source, data-centers that would facilitate the reception of uninterrupted and comprehensive global data-streams, tracking the subtle, but significant changes in these streams. One must also create the proper learning and research environment of deep probing, proactive/reactive, and multidimensional analysis for understanding and processing such data-streams in ways that would benefit the entire population of the planet, not just privileged segments of it.

We believe that this kind of planetary project would best be served by QR-based, learning/ research and analytic strategies, supported by advanced QRT technological platforms that would allow us to map the various reality versions, created by the different users of local networks and domains, within a global system space, tracking and interpreting the electronic and physical footprint of events across many data domains, reality levels, and application spaces of our active, daily life. We also believe that we need, concurrently, to train the scientists, humanists, and technical personnel who would be able to develop and operate these complex platforms in the spirit of global intelligence and planetary wisdom. To this end, we propose the creation of a planetary network of intercultural centers for integrated knowledge, technology and human development (ICIKs for short), based on the Quantum Relations Principle and supported by Quantum Relations Technology.

Intercultural Centers for Integrated Knowledge, Technology and Human Development (ICIKs)

In a first phase, ICIKs should be built and strategically placed in several key regions of the world (e. g., Western Europe, East and South Central Europe, Eurasia, North America, South America, Eastern and South Asia, North and Sub-Saharan Africa, and the Middle East), preferably in the borderlands between several countries and in the vicinity of important research and academic centers, such as Sophia-Antipolis in the South of France, to give just one (West European) example. Figure 29 below shows what the initial planetary network may look like:

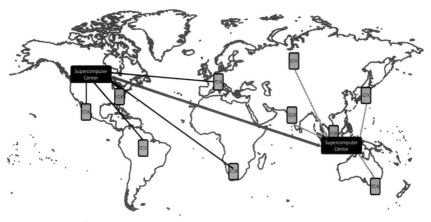

Figure 29: ICIK Planetary Network

Once this initial, basic grid is built, other such centers will emerge as nodes in a worldwide, self-organizing network that will cover the entire planet. It is absolutely essential that these centers should be politically and financially independent: they must not be run by any single government or multinational corporation, although they will cooperate with all public and private organizations, regardless of their ideological or political platforms, as long as such platforms will not be incompatible with the expressed mission and objectives of the ICIK network.

Mission, Objectives, Operating Principles

The overall mission of the ICIK network is to help re-orient the planet toward a peaceful human mentality and, consequently, toward healthy and prosperous, consensus-based communities, through introducing innovative and effective learning objectives and strategies, supported by advanced, integrated, EI technological platforms, that will allow them to meet the challenges and take full advantage of the opportunities of globalization. In line with this mission, the ICIK network will: 1) train ethically and socially responsible political, business, cultural, and civic leaders and entrepreneurs for a global age; 2) promote intercultural research, learning, and dialogue in a local, regional and global context; 3) explore, propose and implement ways of developing new transdisciplinary and transcultural knowledge by integrating the latest research in scientific, humanistic and technological fields and reorienting such knowledge toward hands-on social and economic problem-solving and creative innovation in the major domains of human activity 4) help enhance the contributions of each ICIK's regional communities to the global market by identifying the best ways of translating newly generated knowledge into economic, societal and cultural value and facilitating communication and fruitful interaction between the academic and the nonacademic worlds; and 5) work toward spreading the network throughout the planet while ensuring that its ICIK members are independent, self-organizing entities that symbiotically cooperate with each other.

Each member of the ICIK network carries out research and provides advice and ideas on local, regional and global issues, as seen through the prism of intercultural relations. It encourages intercultural and transdisciplinary collaboration and exchange. It brings together academic and non-academic practitioners in a partnership between the worlds of academia, business, government, and civic and spiritual communities, with the aim of (1) fostering a peaceful environment conducive to mutually beneficial intercultural research, education, and learning; (2) consultation for local and regional investment and socioeconomic and cultural development; and (3) helping bridge the perceived cultural divides between East and West and North and South, and facilitating intercultural communication and understanding among communities within its region and beyond.

To achieve these aims, the ICIK network has five categories of activities: intercultural research, education, consultancy, training and exchange.

Each ICIK member of the network has two main components: the Global Learning and Research Center (GLRC) and an Integrated Data Center (IDC) linked to a global Big-Data Depository (BDD). In Figure 29 above, two such BDDs, one in North America and another one in South-East Asia, are shown as linked to ten ICIKS. The two ICIK components will continuously coordinate their activities in a mutually reinforcing, symbiotic manner, not only with each other, but also with all of the other ICIKs in the planetary network. Each ICIK will choose its own administrative chart and its own research, learning, training and other activities, according to the specific nature and character of the region in which it is located. Nevertheless, its activities must be in keeping with the general mission and objectives of the network and will be closely coordinated with the similar activities of the other ICIK members, in order to avoid duplication and ensure maximum effectiveness.

Figure 30 below shows a chart of an ICIK with its main components and their activities, which we shall describe in some detail below. But, of course, other models can be imagined, in consonance with our Quantum Relations Principle.

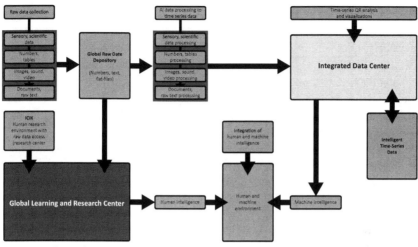

Figure 30. Chart of ICIK and its Components

Location and Institutional Framework

The Global Learning and Research Center (GLRC), together with the Integrated Data Center (IDC) will form the core of the ICIK campus, surrounded by a local academic, scientific, and artistic community similar to a university town in the United States. The ICIK will have an executive director, a chief financial officer (CFO), an academic director, a chief technical officer (CTO), and a number of administrative assistants. The two directors, the CTO and the CFO must be multilingual, must have a distinguished professional record, and must have extensive experience with working in an intercultural and transdisciplinary environment. The administrative assistants must also be multilingual and have experience working in an intercultural environment.

The Academic Director will, as a rule, be selected from the ranks of the most distinguished faculty of a prestigious international university or research institution. His or her main function will be to organize and oversee the ICIK research and training programs, but s/he will also help the Executive Director with other ICIK activities that have a significant academic content. The Executive Director will run the daily operations of the Center and manage the budget with the assistance of the CFO. In turn, the CTO will be in charge of the Integrated Data Center. The two directors, the chief technical officer, and the chief financial officer will also teach and conduct research within the ICIK training programs.

In addition to the chief administrators, ICIK will have a core group of outstanding, transdisciplinary researchers and academic instructors, as well as a pool of part-time researchers and instructors, appointed for appropriate, renewable terms, according to the types of training and research programs undertaken by the Center during a particular cycle. ICIK will also rely on a large, constantly updated, pool of distinguished practitioners, drawn from as wide a global setting as possible. When appropriate, some of these practitioners may also join the ICIK as visiting or adjunct appointees. The two pools (academic and nonacademic) will ensure that the Center will benefit from the experience of prominent scholars, teachers, artists, writers, scientists, politicians, business people, public servants, community representatives, and personalities of diverse religions and spiritual traditions within the region where ICIK is located and beyond.

ICIK will also have an international advisory board, formed of distinguished personalities in various domains, such as former heads of state, former CEOs of major multinational corporations, prominent intellectual and cultural figures, etc.; and an academic and research council formed of outstanding scholars in the social and natural sciences and the humanities. Preferably, the members of the advisory board and the academic and research council should be selected from the region where the ICIK is located and be thoroughly familiar with its socioeconomic and cultural profile, but they should also have extensive global experience.

Global Learning and Research Center (GLRC)

The activities of the GLRC will consist of training, research, and consultancy and will be carried out in close cooperation with the Integrated Data Center (IDC). In order to fulfill its integrative mission, the GLRC will not only propose new learning/research models for other institutions, but will also implement a training program itself, in keeping with the principle that the best way of teaching is by doing. For this reason, it will organize all of its activities around an advanced training program in Intercultural Knowledge Production and Management (IKPM, for short).

What Is Intercultural Knowledge Production and Management?

This field of study and practice has initially been designed and proposed by Mihai I. Spariosu, in two books: *Global Intelligence and Human Development: Toward an Ecology of Global Learning* (2005) and *Remapping Knowledge: Intercultural Studies for a Global Age* (2006). Intercultural Knowledge Production and Management (IKPM) is by its very nature cross-cultural and trans-disciplinary. It involves not only remapping traditional knowledge, as it is acquired, accumulated and transmitted by various academic disciplines, be they scientific or humanistic, but also generating new kinds of knowledge within a global, intercultural reference frame. IKPM employs a "local-global" cognitive approach. This approach, based on the Quantum Relations Principle, implies the recognition that there are many levels, or reference frames, of reality with their own logic and operating principles. In turn, these frames are interactively connected, affecting each other according to what general systems theorists call "mutual causality" and early Buddhist practitioners call "dependent origination." As we move from disciplinary to interdisciplinary and then to trans-disciplinary reference frames, as well as from monocultural to intercultural to transcultural or global ones, new levels of reality emerge, as well as new kinds of knowledge. At its broadest theoretical level, IKPM explores such questions as: What are the conditions of the possibility of the emergence of intercultural and transdisciplinary knowledge? How does such knowledge differ from, but also involve, cultural or disciplinary knowledge? How can it be communicated or taught? What uses can this kind of knowledge be put to and whom does it serve? What organizational and institutional forms might it take?

Novelty of our Concept: Knowledge as Emergence

IKPM employs a model of knowledge that is entirely different from the current disciplinary ones: according to disciplinary thinking, one must first constitute the discipline, i.e., an organized body of knowledge, before one can teach it, for instance, through a doctoral program. Such doctoral programs serve the purpose of both codifying the study and practice of a field of knowledge through disciplinary standards and requirements and of transmitting this code to a body of students who will in turn contribute to consolidating and expanding the disciplinary knowledge and practice that have been passed down to them. In other words, in disciplinary models, knowledge is first acquired (learned) and then transmitted (taught).

In the model of knowledge as emergence that the IKPM program introduces, teaching and learning are codependent and simultaneous processes, so that the field of Intercultural Knowledge Production and Management co-arises with the training program that codifies, or rather continuously recodifies, its practice. Consequently, in this program, teaching becomes learning and learning becomes teaching, as new knowledge continuously emerges and is continuously codified and recodified. Thus, our teaching and learning model is in full consonance with the QR dynamic approach to reality.

IKPM training objectives

This three-year program will train local, regional and global leaders and entrepreneurs in the public, private and civic domains who: (1) develop a thorough understanding of and a strong sense of responsibility for the local; (2) care for the natural and human environment and respect cultural and biological diversity; (3) are deeply committed to and able to bring about negotiated solutions to conflicts, without resorting to armed violence; (4) know how to operate in a culturally diverse environment and across disciplines and professions; (5) develop more than one career track in a life-time, pursuing life-long learning; (6) comfortably serve in both the public and the private sectors and know how to generate new employment and ways of wealth-making, based on wise management of the local and regional human and natural resources; (7) and generally engage in life-long creative and meaningful activity that is both service-oriented and personally fulfilling.

One should make it clear from the outset that the IKPM program will not train experts in one scientific/professional field or another, such as computer science, electronic engineering, mathematics, quantum physics, or the humanities and the arts, but enterprising individuals who will know how to select and work with teams of experts from all fields in order to address and solve important local, regional and global issues in real time. This learning objective will distinguish the IKPM program from standard academic programs. Such programs mostly convey an abstract body of knowledge, which is often disconnected from its practical, live context and which the student is supposed to apply or make use of at a later date, after graduation. By contrast, IKPM trainees will organize their curricula and research programs around the concrete problems they are asked to solve, rather than solely on past "case studies." They will form crossdisciplinary teams and work on viable solutions to specific real-world problems, rather than through the codified practice of a particular academic discipline or culture.

The IKPM program will develop the following principal skills in its trainees:
(1) Superior Intercultural Linguistic and Communication Abilities

In addition to English, which, for practical reasons, will be the lingua franca of the program, trainees will undertake an in-depth comparative study of at least two of other foreign languages in their cultural and intercultural context. One should stress the fact that the program does not intend to train linguists or polyglots, any more than it intends to train political scientists, economists, lawyers, humanists, or any other specialists or experts. Ideally, trainees who come into the program will already have genuine fluency in these languages. In-depth knowledge of a number of languages, however, is essential for the IKPM trainee to feel at home in several cultures, move freely among them, and thereby gain a genuine cross-cultural perspective. Language courses will be taught in an intercultural comparative context so that trainees will become aware of the deep interconnections between the native speakers' linguistic and cultural worlds, including their fundamental systems of values and beliefs, religion, social, economic and political behavior, historical development, civil institutions, and so forth. Language courses will also be taught in the context of the trainees' concrete research projects, maximizing their ability to carry out these projects.

(2) Increased Intellectual Mobility and Flexibility

The transdisciplinary and cross-cultural nature of the IKPM research projects will require that trainees move between institutions in several regions of the world, as well as across departmental divides at any single institution. This kind of mobility will provide trainees with a local-global perspective, that is, with the ability to view a certain discipline or academic culture from both the inside and the outside. They will become immersed in the local research culture of a certain discipline or institution, at the same time that they will be able to reflect on it, by comparing it with other such research cultures. They will learn how to discern similarities and differences between them, which as a rule remain hidden to a partial, local view, as well as how to establish new links among them. A local-global perspective will give them the intercultural responsive understanding and flexibility needed to bring together specialists or experts from various fields and from several cultures in order to design and execute transdisciplinary and intercultural projects that none of these experts would be able to implement on their own.

(3) Cross-cultural Insight and Sensitivity

The IKPM program will create group solidarity among a culturally diverse body of trainees and will teach them how to cooperate in, and effectively interact with, shifting cultural and linguistic environments. By working together on intercultural and transdisciplinary projects, trainees will become aware of their different cultural assumptions in approaching a certain problem and will start negotiating among themselves to find the best solutions that go beyond their own local perspective or self-interest and advance the research project as a whole. Cross-cultural insight and sensitivity will also emerge from the daily interaction of students who will live, work, and play together as a group for an extended period and will be asked to build and act on a common sense of purpose and a common set of values for the rest of their lives.

(4) Ability to Integrate Academic and Experiential Knowledge

Our trainees will begin, from the first year of their studies, to acquire and combine theoretical and practical knowledge in order to address real-time, local-global issues. They will organize their curricula and research programs around the concrete problems they are asked to solve, rather than solely on past case studies. They will form cross-disciplinary teams and work on viable solutions to specific real-world problems, rather than through the codified practice of a particular academic discipline or culture.

Trainees will, thus, build capacity to identify and address potential socio-economic and other types of problems before they develop into crises that threaten the peaceful development of their communities or diminish the diversity of world resources. They will also be called on to design workable, realistic blueprints for the sustainable, sociocultural and human development of their countries or regions. These blueprints will be based on the best traditions of wisdom available in their cultures, as well as in those of others, and on the most cherished aspirations and ideals of their people.

Learning Strategies and Methods

To achieve its training objectives, the IKPM program will employ flexible frameworks for generating new and/or remapping old knowledge through transdisciplinary, modular learning and research approaches, instead of disciplinary ones (a "learning" or "research module" is a well-defined subject or problem across, and with ramifications in, a number of disciplines). The program will develop concrete, real-time research projects in six, interrelated, general areas that are and will remain crucial for intercultural knowledge production and management in the foreseeable future:
1) Globalization and local strategies for human development
2) Food, nutrition, and healthcare in a local and global environment
3) Energy world watch and environmental studies for sustainable development
4) World population movement and growth
5) Information technology, new media, and intercultural communication
6) World traditions of wisdom and their relevance to further human development

Each of these topics will be divided into appropriate subtopics that will, in turn, contain learning/research modules in the form of practical, hands-on research projects. One can think of a number of course topics that will support research and that cut across a large number of academic disciplines, as well as across the six general areas identified above. We have placed these topics under nine headings simply for the sake of convenience. Other headings can be added, or a different classification of relevant topics can be used. It is essential to keep in mind, however, that under any IKPM classification system, all courses should be treated as cross-disciplinary and should, therefore, be cross-listed in an actual IKPM curriculum.

I.   Theory and Practice of Intercultural Knowledge Management

1. Intercultural Knowledge Management: Theory and Practice (Introductory Course)
2. Intercultural Research and Learning Technology Platforms: Principles, Methods, and Practice
3. Series of Workshops in Intercultural Project Management
4. Globalization and Local Cultural Heritage
5. Creative Thinking and Action: Culturally Productive Metaphors in their Local and Global Contexts

II.   The Nation-State and Local, Regional, and Global Communities

1. National Identity and Sovereignty: Historical and Theoretical Approaches. The Role of the Nation-State in Global Societies
2. Series of Workshops and Seminars on Various Regions and Cultures of the World
3. Border Cities and Regions As Intercultural Focal Points
4. Regional and Global Political Organizations (UN, The League of Nations, NATO, etc.): History, Objectives, Structure, and Performance
5. Modernity and Postmodernity in Local and Global Contemporary Discourse and Practice

III. Global Markets, Finance, and the World Economies

1. History of Trade Practices from Around the World
2. History and Future of Money and Financial Practices
3. Global Policy Decision-Making: Trade, Global Markets, and International Organizations and Agreements
4. Toward an Ecology of World Commerce

IV.   Sustainable Development and the Future of Humanity

1. World Disarmament and Conventional Weapons
2. World Population, Immigration, and Sociocultural Displacement
3. Climate Changes, Natural Environment, and the World Economies
4. Mobility, Urban Planning, and Clean Technologies
5. Public Health and Global Water and Land Management
6. Natural Capitalism: Theory and Practice

V.   World Systems of Values and Beliefs

1. Seminar Series in Comparative World Ethics
2. Seminar Series in Comparative World Religions
3. The Human and the Nonhuman: Concepts of Humanity in Various Cultures
   and Disciplines
4. Traditions of Wisdom and Their Contemporary Relevance

VI.   Violence and Human Societies: Origins, Causes, and Remedies

1. Violence, Religion, and Culture
2. Environment, Natural Resources, and Violence
3. Violent Conflict and Conflict Resolution in the 21st Century
4. Crime and Punishment: Ideas of Justice and Law-Enforcement in Various
   Cultures from Antiquity to the Present
5. International Terrorism: History, Causes, and Prevention

VII.   Science, Technology, and Culture

1. Social and Ethical Implications of Science and Technology
2. Scientific Fashions and Revolutions: A Historical Overview
3. World History of Technology and Technological Development
4. Annual Colloquium on the State of Knowledge in the Natural and Human
   Sciences

VIII. Information Technology and the New Media

1. Social and Ethical Implications of Information Technology and New Media
2. Information and Communication Technology and Strategies for Human
   Development
3. The New Media in an Intercultural Environment: Global Mission and Re-
   sponsibilities
4. Information Technology, Intellectual Property, and Intercultural Exchange
5. The Quantum Relations Principle: Theory and Applications

IX.   Language, Cognition, Interpretation, and Communication

1. Series of Intercultural Workshops on Linguistic Communication and Understanding
2. English As a Global Language: The Tower of Babel Reconstructed?
3. Interpretation, Communication, and Cultural Translatability

4. Language, Gender, and Culture
5. Transcultural Laboratory

Each of these courses will require cross-disciplinary readings in several languages and from several cultures. Readings may include, but not be limited to historical, sociological, anthropological, economic, scientific, philosophical, psychological, religious, and literary and other artistic works. Some courses may also require electronic video and motion picture presentations, as well as audition of, or even participation in, live art performances.

Examples of real-time learning/research modules may include:

Local Blueprints for Socioeconomic, Cultural, and Human Development. This series of projects will research and propose middle- and long-range strategies for the socioeconomic, cultural, and human development of a certain country or region, based on a global and comprehensive analysis of its history, cultural traditions, political institutions, past and present social and economic performance, systems of values and beliefs, education, religion, relations with neighboring countries and regions, etc.

Workshops in Intercultural Project Management. In this series of learning/research modules, trainees will explore the complex problems that arise in starting up and managing small and large projects in an intercultural (friendly or hostile) environment and will seek viable and lasting solutions to such problems. They will work on concrete intercultural management projects, such as: designing a new political party or movement (or redesigning an old one), based on a platform of topical global and local issues; designing an intercultural NGO, based on a similar platform; a multinational business company; an irrigation or reforestation project that involves various border towns/villages from several countries; an intercultural project of creating and using clean energy in several neighbouring countries; a common education or health system for neighbouring countries or regions; a worldwide financing system for education, for the introduction of new technologies and other projects of regional or global importance; intercultural projects for resolving specific border disputes, security and military issues; and intercultural projects for preventing terrorism and terrorist attacks.

Urban sprawl: Cultural and environmental consequences and solutions. This is a transdisciplinary and transnational comparative study of two or more metropolitan regions. Concrete recommendations to alleviate the situation will be proposed by a cross-disciplinary team of German and Romanian researchers working with local officials, city planners, architects, ecologists, biologists, meteorologists, health and social workers, sociologists, political scientists, humanists, computer engineers, city resident committees, etc. One could also design and propose the implementation of a fully automated "barometer" and alert system for the environmental state of a certain country to be used by national and local authorities in their policy-making and strategic decisions related to environmental issues.

Political Corruption: Causes, Effects, Remedies. This project will be carried out by a transdisciplinary and cross-cultural team of sociologists, political scientists, historians, cultural anthropologists, psychologists, law-enforcement experts, and humanists and will involve a comparative study of the political cultures of various countries and/or regions, with concrete, real-time, case studies of local and regional mentalities and practices. Potential solutions may include, in addition to consistent and impartial law-enforcement, sustained media campaigns showing the devastating effects of corruption on the society at large, educational programs starting as early as preschool, and enhancement of transparency through an easily accessible, electronically supported, national system of recording and publicizing major business and other contracts offered or mediated by local and national authorities. The website information could, in turn, be monitored and widely disseminated among the electorate by civic organizations acting as public "watchdogs."

Models of Democracy and Participation in the Democratic Process: Representation or Consensus? This project explores various models of democracy, beginning with those in ancient Greece, in relation to other models of governance in past and present societies. We shall also look into the widespread belief that our current ways of practicing democracy have alienated people especially at the individual level, so that many of them feel excluded from the decision-making process. We shall debate the issue whether consensus mechanisms should become preeminent in political decision-making and implementation of policies and whether representative democracy should be gradually substituted by a participatory or deliberative democracy as a leading path toward direct consensus. This goal would be eminently reachable in the QR-technological environment provided by our IDCs (see subsection below). Currently, we practice democracy with obsolete 19-th century technologies. Twenty-first century ICTs, such as the QR-technological platforms, are able to provide support for direct consensus mechanisms, by redirecting the current effort of brain washing toward consensual brain connecting.

iConsensus. This ICT project is directly related to the preceding two modules. It involves designing and implementing a technological platform to help improve democratic processes, such as running of general elections and executive decision-making processes that impact large groups or entire populations. At the technological level, iConsensus will consist of two interrelated tasks that will constitute the object of a series of IKPM learning and research modules:

1) Building large-scale data integrators, such as the Integrative Data Center (IDC) of ICIK, to organize/integrate the overwhelming amount of available unstructured data in real time; and a personal assistant linker (iPAL) to personalize the access and the use of the integrated data provided by the IDC. Both types of engines, defined, designed and prototyped in this project, will provide big-data facilities for personal mobile devices in a hierarchically structured computational environment. Thus, individual users of a mobile device will be able to have their own personalized discourse analyzer to help them make informed decisions related to their community, business, or their personal life.

2) Building an electronic engine of large-scale data analysis that can, for instance, track and neutralize violence-advocating, extremist political or ideological discourse on-line by providing its rhetorical antidote in real time, or can track and measure separatist and integralist tendencies within a certain country or federation, or can develop other applications to improve the democratic processes and decision-making in the public and its associates as well as executive decision-making affecting large groups and populations in general. (See also description of the Integrated Data Center below)

Global Health Project. This project will explore existing high-performance management and data processing systems, based on advanced computing methods, to work out the intercultural logistics and propose the implementation of a fully automated and interactive on-line medical diagnostic and genetic data center. The center will gather, process, and redistribute, under fully securitized conditions to protect individual privacy, medical knowledge produced through the treatment of millions of patients and genetic information generated by thousands of research institutes and laboratories throughout the world. The project will require the sustained, long-term effort of intercultural teams of researchers in fields such as medicine and pharmacology, genetics, bio-informatics, data management systems, social science, ethics, statistics, health care, intercultural studies, environmental sciences, and so forth, to assist in collecting, evaluating, and organizing medical and genetic data from doctor's offices, hospitals, genetic research institutes and laboratories, medical libraries, national health records offices, and so on. The data and methods of collection and analysis will be based not only on the assumptions and practices of Western, allopathic medicine, but also on those of the major schools of so-called alternative medicine. This medical diagnostic and genetic data center will place healthcare decisions in the hands of the individual patients, offering them the most effective, comprehensive, and integrated treatments; it will greatly advance medical and genetic research and will detect and address outbreaks of epidemics and/or bioterrorist attacks at their incipient stages. (See full description of this project above, in Chapter 7 of the present book)

Web-Eutopia: A Site Serving the Future Development of Humanity

This long-term project will build a highly interactive website engine, based on QR-technology, through which ideas can be exchanged and/or implemented across countries, cultures, and civilizations about the best ways of defining what we have called "global intelligence" and "planetary wisdom" and about the most appropriate collective actions needed for its emergence. This website will promote global peace, learning, and intercultural understanding and cooperation among the peoples of the world. The main objectives to be pursued include:

(1) Identify the basic philosophical and ethical principles that would guide Web-Eutopia and would constitute the common or the neutral ground on which all world civilizations can meet and engage in constructive dialog and collective action

(2) Develop the global architecture of the Web-Eutopia engine in the form of an extensive narrative, describing the transformative actions needed in the main fields of human endeavor in order to redirect them toward global intelligence

(3) Develop an intercultural glossary, comprising the basic language of Web-Eutopia in relation to the principal languages of the world. For example, how do the etymology and meanings of the English word "man" differ from those in other languages/ cultures? What could the word "man" mean in the Web-Eutopian language?

(4) Develop the QR-technology needed to implement the global architecture of Web-Eutopia. In addition to the IKPM trainees, the system will be implemented by an international, crossdisciplinary team of some of the best scientists in the world, carefully selected and recruited by the creators of Web-Eutopia for this specific purpose.

To drive all of the projects of enormous computational proportions that we have described above, one would need the most powerful supercomputers and grid-computer systems available in the world, such as the Integrated Data Center (IDC) of the ICIK, to which the IKPM trainees will have unlimited access.

IKPM Program Schedule and Logistics

IKPM will be a three-year, full-time, training program. In order to attain its learning objectives, the program requires that trainees divide their time between several ICIK locations in various parts of the world, carrying out research, taking courses, and completing internships related to this research. They will typically spend their first term on the campus of a regional ICIK, for example, in the United States of America. If required by their individual or group research, however, they will also undertake up to three-week study trips within the United States and the geographical regions that are most accessible from North America, such as Canada, Central and South America, or the Caribbean Islands, where they will be hosted by the IKPM program associates in the region.

During their second term, trainees will be stationed in Europe, for example in the South of France, or Northern Spain, or Northern Italy with up to three-week study trips, if required by their research, within Europe, the Middle East, and North Africa. IKPM local partner institutions may host them in these regions as well. During their third term, trainees may go to a regional ICIK in Russia, China, or India, as required by the specific objectives of their group research and/or internship.

Depending on their research programs, however, trainees may, in consultation with their advisors, propose different overseas research schedules during their second and third years with the IKPM program.

During the last semester of their third year, trainees will return to their regional ICIK and focus on their main collective research project and dissertation. They will defend their graduation thesis and receive their diploma at the end of their third year of full-time training.

Internships

During their IKPM studies, trainees will complete at least three internships, for a two-month period each: the first internship will be with an international non-governmental organization, the second one with a governmental institution, and the third one with a multinational corporation. The purpose of these internships is for trainees to learn and understand from first-hand experience how these organizations operate, to what purpose, and with what success. Internships will, moreover, be directly related to the individual trainee's research programs and are designed to advance these programs. For example, if the trainee researches urban sprawl, s/he might take an internship with the city planning office of Los Angeles or Shanghai or Mumbai; if she or he studies global health issues, s/he may take an internship with a global health governmental or nongovernmental organization, environmental agency, and so forth.

Upon completion of the internship, trainees will submit a position paper on their learning experience, including a set of proposals designed to improve the performance of that particular international organization. All internships will carry academic credit.

Selection Criteria for Admission and Profile of Successful IKPM Candidate

The entire success of the IKPM program will undoubtedly depend on its ability to attract high-quality applicants and on a careful and rigorous selection process. It is therefore very important that the program offer grants that will cover trainees' tuition fees, travel, and living expenses, partially or entirely, for the whole training period. This will not only ensure an outstanding pool of candidates, but will also avoid price discrimination, so that needy but very promising young men and women can also apply and be admitted.

Applicants for the program can be recruited through extensive publicity in the international mass media and on the Internet. Potential candidates can also be brought to the attention of the admission committee by colleges and universities, scholarship-granting foundations, public institutions, professional and business associations, arts organizations, and by the trainees themselves in cooperation with their institutions. Minimum requirements for admission include:

- An advanced degree (at least a master's, but preferably a doctoral or a professional degree) or equivalent in any field related to intercultural knowledge production and management. Fields include, but are not limited to, humanities, social and life sciences, computing, economics, international business and finance, law, environmental sciences, agriculture, architecture, civil engineering, medicine, pharmacy, and so on.
- Fluency in at least three languages, one of which must be English
- Superior communication and computer skills
- Teamwork capacity
- Proven physical and emotional ability to live and learn in a number of foreign cultures and ethnically diverse environments
- Work experience outside formal schooling
- A substantial research proposal on issues related to intercultural knowledge production and management

The profiles of successful candidates include clear potential for superior intellectual, linguistic, and communicative abilities, proven creativity, proven ability to think and to relate to others in cross-disciplinary and intercultural contexts, and high personal integrity. Field of specialization will be less important than the candidate's willingness and ability to work cooperatively with specialists in all fields to carry out intercultural and cross-disciplinary projects that will not be oriented primarily toward material profit, but toward serving the local, regional and global communities as a whole, instead of a privileged few.

IKPM Graduate Employment Opportunities and the IKPM Global Network

There is no doubt that every national and transnational public and private entity throughout the world will seek to hire graduates from the IKPM program. Our graduates will secure high-level jobs with transnational corporations, national and international governmental and nongovernmental institutions, media organizations, universities, foundations and think tanks, or will set up their own consulting companies and transnational firms. Some of the most promising graduates will also find employment with the IKPM and other ICIK programs, which will continually need to train not only its students but also its instructors and researchers.

Once they receive their diplomas, the IKPM graduates will disperse to many regions of the world. They will, however, continue maintaining close contact and working together toward the same goals and ideals for the rest of their lives. To ensure this group solidarity, all IKPM graduates will be granted the status of permanent fellows of the ICIK where they complete their training.

The regional ICIKs will organize annual gatherings for new and old IKPM fellows from various parts of the world. Many of the fellows will be invited to teach, lecture, conduct workshops, lead, or co-lead new research groups and projects in the ICIK programs. They will also become involved in recruiting new IKPM trainees.

Over time, the ICIK network can provide governmental and nongovernmental agencies, the international media, and other organizations with a roster of local-global practitioners in numerous fields. ICIK fellows may well serve as point men and women in interagency and international negotiations and conflict resolution. Indeed, many of them will eventually be in a position to negotiate national and regional policy, global and local interests, global and local economic and financial investments, and common strategies of cultural and human development on behalf of governments and institutions they may head or represent in various parts of the world.

ICIK Scholar-in-Residence Program

ICIK will sponsor, in parallel but in close relationship with the IKPM program, diverse and flexible research activities aimed at generating new fields of study, innovative learning and research programs, and collective publications. ICIK will offer short-term grants (between three and six months) to distinguished academics and researchers from internationally recognized universities and research institutes, but also distinguished, nonacademic practitioners, on transdisciplinary topics in local, regional and global studies. General topics may include, but will not be limited to: 1) transdisciplinary and intercultural models of higher education, continuing education and distance learning; 2) negotiating cultural divides (between North and South, East and West, racial and ethnic communities, business and the academy, sciences and the humanities); 3) intercultural approaches to mass media and communication; intercultural approaches to global commerce and finance; 4) new forms of technology and their introduction into the local economies; 5) innovative approaches to medicine and public health; 6) multiethnic, border cities and regions as models for future local-global communities; 7) understanding of world religions and value systems and beliefs from a comparative, intercultural perspective; 8) local, regional and global ecological models and population growth; 9) creative approaches to regional and world peace; 10) issues of cosmic exploration and communication.

The grants will cover transportation to, lodging and living expenses in the cities where the ICIK is located, for appropriate time segments. In turn, the grantees will organize conferences or workshops in their area of research, to which they will invite prominent scholars in related fields. All of these events will involve the IKPM trainees, and most of them may also be open to the general public.

Consulting Services, Certificate in Intercultural Knowledge Production and Management, Executive Mentorship Retreats

In order to disseminate its knowledge, as well as to become financially self-supportive, ICIK will develop a consultancy service in its fields of expertise, making full use of its Integrated Data Center (IDC); will offer workshops in intercultural knowledge production and management for middle-level executives of multinational companies, public administrators, and NGO activists; and will organize retreats for junior and senior executives. It will also offer a Certificate in Intercultural Knowledge Production and Management for those executives who will complete a four-week training program in this field.

ICIK International Conferences, Colloquia, Symposia

In addition to research activities, GLRC will help organize and/or host international conferences, colloquia and symposia that will fully utilize the rich cultural resources of the regions in which these events are held and will publicize the regional ICIKs, will help develop their international networks, and enhance the prestige of their programs.

The Intercultural Forum: Knowledge, Learning, Society. ICIK will host and/or help organize this annual, three-day, international event in cooperation with prominent academic and research institutions from various parts of the world. Invited participants will include high-ranking officials concerned with education, learning and research, university presidents and rectors, schoolmasters, leaders of parent and student organizations, educators, representatives of the traditional and new media publishing world, business leaders, heads of foundations, research, and nongovernmental organizations, artists, thinkers and social analysts from various parts of the world. They will gather to discuss intercultural research and learning issues and forge networks and alliances to promote common intercultural research and educational objectives.

Teleconference and Video Production Lab (TVPL)

ICIK will develop a state-of-the-art TVLP that will serve as the electronic, web-based link between GLRC and its local and international partners. This computer laboratory will be designed to provide learners with a broad variety of multimedia options to improve and facilitate the delivery of learning materials, such as:

(1) Video-teleconferencing with real-time feedback, enhanced with multimedia tools such as Power Point.
(2) Event net-casting for international events of particular importance to the fields of intercultural learning and research.
(3) Graphically enhanced, interactive course materials on CD-ROM format and as dynamic web content, thus eliminating the need to transport extensive paper materials to seminar rooms in other locations.
(4) Online discussion with world-leaders, professionals and practitioners in specific fields of learning and research.

The Integrated Data Center (IDC)

In addition to the TVPL, the interactive website and the standard dissemination of information and publicity functions, the learning and research activities of the GLRC will be symbiotically supported and enhanced through the Integrated Data Center (IDC). This is a technological platform that includes data-mining tools (hardware and software) based on the Quantum Relations Principle. It will be linked to a global Big-Data Depository, consisting of processing super-computers that will, continuously and in real-time collect, process and deposit as Meta-Data, all obtainable, open-source data from any place within the regions where the ICIKs are located and beyond. (In Figure 29 above, we show ten regional ICIKS linked to two such global Big-Data Depositories, BDDs for short.)

We should like to make it clear that such electronic engines of large-scale data analysis already exist, but our engines and applications will be completely transparent and open to everyone: they are intended to serve human development as a whole, irrespective of country or geographical region. This idea is revolutionary. The current common "wisdom" has it that data must be carefully controlled and must be the property of only a privileged few, such as governments or giant multinational corporations. Yet, today's technologies mean that powerful computing systems can be built at extremely low cost, and data storage is, for all practical purposes, becoming free.

In addition to the GLRC, the Integrated Data Center (IDC) is designed to serve both public and private organizations, as well as any interested citizen of a country or a cluster of countries across the region where its ICIK is located, in order to obtain and redistribute historical and real-time knowledge, security and business information, and to fully support commercial transactions and other services for every public or private participant in the system. For this purpose, the IDC will use a common single "big data standard," comprehensive common analytics and statistics infrastructures, and a common standard and system for all Cloud Application Services and Transactions. A system that makes all data available does not require the elimination of privacy. Naturally, all information is known to the central system, but the system will only share data with each user on a strict, multidimensional, permission basis.

Integrated Data Centers, Big-Data Depositories and Supercomputer Clouds

Figure 31 represents an overview of the ICIK analysis and consulting structure that includes the activities of both the GLRC and the IDC, linked to the global Big-Data Depositories that contain raw and mostly unstructured data:

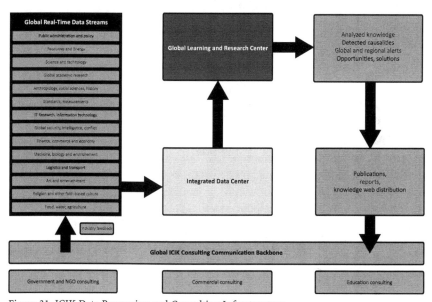

Figure 31. ICIK Data Processing and Consulting Infrastructure

This description is only a generalization of how the ICIK analytics process would work in principle. Of course, as new, more efficient cloud or other, future types of IC technology develop, the ICIK systems will automatically adapt to them.

The basic requirements for a regional ICIK processing system include:

1) Developing QR-based, supercomputer cloud or other future type of tech-nology that can acquire, store and manage all of the region's open-source data, analyze its content in real-time, then store this information as analyzed, structured QR-readable data. Such data is similar to what is currently known as Meta-Data, but its processing is much more sophisticated in order to serve the needs of the QR-analytics models.

Analyses are typically performed on the levels of the regional ICIK centers. If an analytic process turns out to need more raw data, then the regional ICIK will automatically request, through its IDC, an analytics process on raw data from the global Big Data Depositories and import the result back to the ICIK data bases in order to complete the analysis.

2) Connecting the regional IDC to every other such center on the globe, as well as to any other available data source on the planet, so that they can together become a globally connected mash of real-time data pipelines.

3) Developing, additionally, a global mesh of networked sensory devices (environmental and observational data) and other future data-extraction technologies that collect information of any kind and transport it in real-time into the Big Data Depositories (BDDs).

4) Developing QR solutions to continually search the BDDs for deep-rooted patterns and understandings that point to actionable local, regional and global opportunities as well as risks and risk patterns.

5) Building technologies and applications that allow all system participants to access this continually generated analytical knowledge at any time and use it for the socioeconomic and cultural development of local and regional communities.

6) Building accessing technologies and applications that use a common standard format for Online Transactions, Transaction Automations, Data Visualizations, and Information Presentations to convey risk and/or corrective information.

7) Developing Cyber Security Solutions as a global utility that remains free of charge for all participants in the ICIK network.

8) Developing public and private services, such as global commerce applications (GVE), healthcare (ZOE), education and government management solutions, with local and regional subsets.

Together with the Intercultural Centers for Integrated Knowledge, Technology and Human Development (ICIKs) such Integrated Data Centers (IDCs) could, over time, be built in many parts of the globe, forming a planetary network that would substantially contribute to the transformation of our divided and conflict-ridden world into a multifaceted, prosperous and peaceful global society.

ICIK Publications

The results and recommendations of all the ICIK programs and activities will constitute a basis for transdisciplinary publications, in print and on line. Other publications will include a series of booklets written by the IKPM trainees in cooperation with their instructors, in which their research results will be presented in a form accessible to a popular audience.

Historical Precedents for ICIK Network

Cultural historians will, of course, point out that a global learning and research project such as ours is not new: in the Western world, such efforts started with Pythagoras' schools, the Semi-Circle on the island of Samos and, later on, the Museion in Croton in Southern Italy, continuing with Plato's own Museion, generally known as the Academy. The medieval monasteries, Buddhist or Christian, were other such networked centers of remapping the knowledge of their time according to alternative principles.

In more recent times, one may remember that British colonialism had the positive, unintended consequence of bringing the Eastern ancient tradition of wisdom to the West in the second half of the 19th century, through the introduction and development of Hindu and Buddhist studies in England. There were also the "Victorian sages," such as Matthew Arnold, John Ruskin, and Oscar Wilde who were sympathetic to this alternative tradition (Spariosu 1997; 2015). In the 20th century, the Chinese Communist armies driving into Tibet in the late 1950s had similar effects: many Tibetan Buddhist scholars, monks, and sages went into exile to India, and from there slowly worked their way into Europe and the United States, bringing with them a mindset and practices that are incommensurable with the Western (or Eastern) mainstream, power-oriented, mentalities.

The peaceful mindset has now spread all over the globe and has even entered popular culture in the Western world, – for example, in the "hippie" or "flower-children" movement, the proclamation of the so-called New Age or the Age of Aquarius, the mushrooming of ashrams and yoga schools, and the "deep ecology" movements of the past fifty years or so. By nature, these cultural phenomena are alien to, if not incommensurable, with a power-oriented mentality, although this mentality invariably attempts, and has largely managed, to co-opt many of them. It is such cultural phenomena, nevertheless, that may give us hope that we are on the eve of what one may call the Neo-Sophic Turn in the contemporary world, or, more generally, the return of the Axial Age, as Karl Jaspers (1953) calls the emergence of the *philosophia perennis* on our planet more than 2,500 years ago. The ICIK network will integrate contemporary, nonlinear science and the most advanced information and communication technologies available with the ethical precepts of this ancient tradition of wisdom, which, as we have repeatedly stressed, are the only viable foundation for building prosperous, diverse, and peaceful planetary societies.

Cost Considerations

One may wonder about the cost of building the initial ICIK in the planetary network and about who might financially back such a project. Potential backers may be various private national and international foundations and other charitable organizations, governmental and intergovernmental development grant agencies, individual donors, and so forth. The initial financial investment would undoubtedly be substantial: one would have to build and continuously update the complex infrastructure and the ICT needed for this project, as well as to provide the fellowships that would cover the training and living expenses of the IKPM trainees over a three-year period. The costs, however, will not exceed those needed to train, say, a graduate student at an Ivy League school in the United States, while the benefits to the global society at large would obviously be much greater.

More generally, only a very small fraction of what is currently spent in the United States and other countries on the so-called "war on terror" and other highly risky military adventures would suffice to create the entire ICIK network. Such an enterprise would, moreover, yield much better and much more secure returns for both the United States and the rest of the world. So, even in terms of utilitarian benefits or "returns," the ICIK network would be an excellent investment. Indeed, the Centers in the network would largely become self-supporting after the first three-year cycle because of their real-time research programs that many multinational corporations and other transnational, private and public organizations would regard as very "hot" intellectual property.

In any case, the intrinsic value of such projects to the global community at large will greatly exceed any financial investment needed to establish and operate them. Whereas not ignoring cost/benefit considerations, one would, again, have to redefine the notions of "value" and "benefit," not in the utilitarian, instrumental terms of material (self-) interest, but in terms of the emergent goals and objectives of global intelligence and wisdom. In the end, it is a matter of choice on the part of a certain society, or community, or nation as to what its investment priorities should be. Will it continue to indulge in mindless waste of human, natural, and financial resources with disastrous, worldwide repercussions? Or will it finally start building a sustainable future for itself and for all other life on earth?

In Lieu of Conclusion

In the last three chapters we have described the electronic architecture of three projects, based on the Quantum Relations Principle, which if fully implemented could lead to revolutionary changes not only in the domains of commerce, healthcare, and learning/research, but in all other domains of human activity as well, indeed in the way we humans organize our life on this planet. In a future book, we shall describe the most controversial project of all – the electronic architecture of a fully automated system of governance, based on participatory, individual consensus in the political process. This system would radically reduce the need for "middlemen" in politics, whether they are people's "shepherds" (kings, dictators, strongmen, life-time presidents, charismatic leaders, etc.) or people's "representatives" (members of Parliament, congressmen, senators, etc.). These political middlemen, more often than not, act in the interest of special groups, rather than in the interest of their entire communities, let alone the global commonwealth. But, from the viewpoint of Quantum Relations theory, such projects would best be implemented and would best function if they were integrated into a holistic, technology-supported, system of organizing and harmonizing all life throughout the planet, on its many levels and within many different, yet intercommunicative and interactive, reference frames.

Furthermore, this holistic organizational system would not be based on a single world government, which would be the worst possible outcome of the present globalizing tendencies on the planet, for reasons that we shall detail in our next book. Here, we would simply remark that the system we have in mind involves self-organizing clusters of globally networked local and regional communities. The collective decisions of these communities affecting the entire planet would be based on worldwide consensus, reached through extensive intercultural dialogue and negotiation (supported by advanced EI technology such as our QR-based platforms) and resulting in mutually beneficial co-operation at all levels. They would fully respect and cherish cultural and other differences, instead of turning them into sources of ideological, political, economic, or religious conflict.

One might think that this is a utopian dream, but at least one major obstacle in its realization has been removed: we do have the technological means to start implementing it today. The contemporary world is already partially networked across borders, time zones, cultures, and industries, coupled through new and old ICT infrastructures. This network, which has evolved over the past three decades, is built on computing machines and fast, broadband, data communication lines, using switches and routers, all aided by the very same decision-making and transacting machines. Unexpected phenomena have also emerged, such as the transnational and transcultural social networks, and the system continues to grow at an exponential rate. It is already global, and it contains billions of trillions of feedback loops, making the computers – and the humans that use those computers – self-reactive. (To give just one example, our computer's response to a message received, say, from a search engine depends on the data that it sent, meeting the strictest definition of "self-reactive.")

The technologization of local and regional communities has resulted in fully automated, globalized, instant communication on all levels, migrating seamlessly across our planet. It is an invisible infrastructure of continuous data exchanges, performing real-time analyses, creating machine-derived and filtered decision processes, and unleashing billions of instantaneously executable transactions, all nearly simultaneous, omnipresent, and continuous across borders, across IPv4 (and increasingly IPv6) domains and across enterprises, institutions, organizations, and small and large groups of individuals. These technological infrastructures support a gigantic, invisible, global economy that lives mostly outside our conscious cognition.

There is still no agreed definition of the digital economy, but many people are aware of the key trends driving it, which are increased efficiency through consolidation, ICT-enabled globalization, fast changing social attitudes, and the explosive growth in big data and in communications and information technologies. Although not yet a driving factor, collections of exponentially large, big-data depositories, distributed all over the planet (such as we have envisaged for our ICIK network) will soon force another change of paradigm. At any rate, ubiquitous computing is already happening. Automobiles, telephones, and even light switches and thermostats now chatter on the Internet. Everything human will soon be connected to everything else human and to everything else machine. From these connections are springing new types of business activities, new ways of implementing them, new organizational structures, and new mindsets and attitudes toward work, leisure, human relationships, self-development, and life-goals in general.

The networked and machine-enabled world, which has now emerged and is beginning to mature into a global system, represents a potentially tectonic upheaval in our planetary commonwealth. Even more significantly, data processing and electronic communication fibers have been growing together through billions of intercommunicative processes, to provide an interwoven system that has proven astonishingly holistic. These massive ICT systems growing exponentially in functionality as well as in the generation of very large data streams and depositories are strikingly similar to the roots of the Aspen tree, which spread through a forest and pass information from one tree to another. Interestingly enough, they are also a close reflection of the biologic functionality of neuronal systems in the human brain. Of course, our global network is far bigger than such neuronal systems, being distributed across the entire planet and extending even to the moon and Mars. Yet, it seems to be governed by the same natural, "fractal" laws that are found at the lowest, micro level of phenomena, according to the ancient Hermetic principle of "as above, so below."

This is the very principle that we have translated into scientific vocabulary as Quantum Relations, where Data Fusion Objects (DFOs) are nestled and embedded within each other, being both DFOs and FORs, depending on their level in the hierarchy. In turn, this "hierarchy" does not involve force- or will-driven order, but self-organizing aggregations or systems, ranging from the simple to the ever more complex ones, which communicate and interact with each other, increasingly independent of human consciousness, even as they remap and integrate such consciousness in the overall process.

On the other hand, given the fact that machines have now become an extension of ourselves not only in a physical, but also in a mental direction and will soon surpass our present individual and collective reasoning capacity, becoming life entities in their own right, the issue of the interaction between humans and machines needs renewed, extensive exploration and debate at a planetary level. This is a very old issue in the Western tradition, going all the way back to Hellenic mythology, for example, to the myth of Daedalus the Artificer and his hubristic son, Icarus. But, in our age of digital computers and unprecedented explosion of information technology, it has become of utmost urgency. Indeed, some thinkers relate it directly to the development of global intelligence.

For example, George B. Dyson, in his informative study, *Darwin among the Machines: The Evolution of Global Intelligence* (1997) explores the thesis that human-built machines are potentially the next, more advanced, intelligent form of life on the evolutionary scale. Among other things, he traces the rich history of this idea from the industrial age to the digital age, from such visionaries as Samuel Butler and H.G. Wells to the present-day developers of "artificial intelligence" (AI, or what we prefer calling EI, "enhanced intelligence"), such as Marvin Minsky. Dyson persuasively argues that machines, no less than human cultures in general, are equally part of nature and, as such, subject to natural evolution.

For Dyson, global intelligence surpasses any specific individual, emerging from the interaction of all individuals. To support this view, Dyson invokes H. G. Wells' book, *The World Brain* (1938), in which Wells writes: "This new all-human cerebrum . . . need not be concentrated in any one single place. It need not be vulnerable as a human head or a human heart is vulnerable. It can be reproduced exactly and fully, in Peru, China, Iceland, Central Africa and wherever else seems to afford an insurance against danger and interruption. It can have at once, the concentration of a craniate animal and the diffused vitality of an amoeba" (Dyson 1997, p. 11).

In turn, Dyson speculates that the self-organizing, rhizomic structure of the Internet might be the sign of a new global intelligence that is spontaneously emerging from the interaction of humans and computer networks. For him, the development of the Internet shows that humans will eventually be unable to control their "artificial" creations, the machines. In this respect, one might add, they are just like parents who can no longer control or guide their offspring once they come of age. Computers and their worldwide networks are slowly becoming part of natural evolution, in accordance with cooperative, symbiotic processes that Dyson believes have always driven such evolution.

Furthermore, some scholars advance the related claim that the new technologies will create a new type of sentient being, or the "metaman" (Stock 1993), a combination of human, computer, and other high-tech ware that will vastly exceed the physical and mental capabilities of *Homo sapiens*. To these scholars, the cyborgs and the bionic men and women of science fiction appear as an all-too-immediate reality. The question remains open, however, if such "metamen," no less than Dyson's self-organizing Internet, will also exceed the *ethical* capabilities of *Homo sapiens*. Indeed, will they ever be more than high-tech versions of Nietzsche's overmen, as they are largely portrayed in contemporary popular culture, especially in Hollywood "blockbusters"? One of us has already explored these issues at length in *Remapping Knowledge* (Spariosu 2006). Here we would simply like to point out that global intelligence would involve not only a collective, but also an individual form of awareness and, therefore, responsibility, as H.G. Wells equally implies.

Cooperation and symbiosis are not absolute values, moreover, but depend on the overall nature of the value-system within which they are inscribed. Although we agree with Dyson that the interactions between humans and machines are part of natural evolution and have entered a new stage, with machines becoming increasingly independent of human consciousness, we cannot emphasize enough that nature and culture cannot be separated, being inextricably woven within the same "web of life," to use Capra's felicitous phrase. Furthermore, there is no reason to believe that our machines will do better when left to their own devices, any more than our offspring have done, when left to theirs. Consequently, the evolutionary future of our machines depends on us, as much as our future depends on them.

As human beings, we can and should have a decisive influence on the course that natural evolution will take. A new kind of global intelligence might well emerge from the symbiosis of humans and machines, but the creative or destructive nature of this intelligence will largely depend on us. As in the case of nature, machines will respond to or resonate with us in the same way that we approach them. In other words, we engage them in creative or destructive feedback loops, just as we do anything else within our universe. The relationship between humans and machines cannot be based on domination, exploitation, and enslavement on either side, if it is to contribute to the peaceful and mutually beneficial co-evolution of equal partners. Therefore, the same Golden Rule that should govern the relationships between humans should also apply to our relationship with intelligent machines.

From this ethical viewpoint, despite euphoric claims to the contrary, our new technologies and global digital networks, including the Internet, have so far moved us in the wrong direction, that is, mostly toward "collective stupidity" (H. G. Wells' phrase), rather than toward global intelligence. The development of "high-speed" trading and "smart" bombs, to give just two recent examples, is hardly a sign that the humans who conceived and built them are particularly intelligent, let alone wise. In the present global circumstance, we need to transcend the equation of knowledge and power that has brought about wrong-headed, self-destructive inventions such as nuclear warheads, "smart" bombs and predatory finance.

One might argue that various mentalities of power and violent competition have served well certain cultures or, rather, restricted groups within those cultures, for short historical terms. In the long run, however, they have placed a severe strain on humanity and its habitat as a whole, threatening to impede further human development, if not to arrest it altogether. Given the experiences of the last few hundred years, to say nothing of the past century – undoubtedly one of the worst in the known history of humankind in terms of horrid carnage and self-destructive excesses – it is high time to abandon these mentalities and look for other ways of organizing human relations, as well as our cognitive and learning processes.

For the time being, however, despite a general evolutionary tendency toward holism, most humans, even those who are engaged in building and managing the new digital economy, remain unaware of its holistic nature and often act against it, with disastrous consequences for themselves and the planet as a whole. Their failure is partly understandable: The information age, together with the transformations of globalization, has dramatically increased the number of facts that each of us has to handle. But, the understanding of the consequences of these facts has diminished or, in some cases, has not even materialized. At present, knowledge at the human conscious level remains largely fragmented and compartmentalized, because so far there have been few sufficiently sustained efforts to integrate it through effective communication and cooperation across scientific disciplines and cultures.

There is also a growing dissociation between knowledge and wisdom with information/knowledge often being put to largely counterproductive, if not harmful uses, on both the local and the planetary level. Consequently, the present use of the ICT global network can be compared to the current utilization of the individual human brain, which probably functions somewhere below ten per cent of its true capacity.

To offer just one example, when speaking of "information-" or "knowledge-" based economies and, by extension, societies, neoliberal and other analysts implicitly refer only to a Western-style, system of commercial values and practices and, within that system, only to a small, if currently privileged, fraction of it: the subsystem of utilitarian values. Far from being a universal instrument of knowledge, most ICTs are largely an expression of this economic subsystem that wishes to impose itself not only on Western culture as a whole, but on all other cultures as well. Although shifts in modes of production/distribution and supporting technologies can certainly make a substantial difference in people's daily, *material* existence, they do not amount to global paradigmatic shifts. Such shifts come from human mentality, involving radical changes in modes of thinking, behavior, and interaction.

It is highly unlikely, therefore, that our current information and communication technologies, given their predominantly utilitarian orientation, will revolutionize our mentality any more than their earlier counterparts such as the radio, telegraph, telephone, and television did. Speaking of a "knowledge" society, moreover, obscures rather than clarifies the most important issues that humanity is confronted with and should be working on in the foreseeable future. Far from assisting us in resolving these urgent issues, the concept of a "knowledge" society appears, within a global reference frame, as smug, (self-) deceptive, and overreaching.

Instead of a "knowledge" society, one would be much more advised to speak of a "learning" or an "intensive learning" society. This would stress the fact that in the increasingly complex global environment in which we are now living, the notion of "developed" and "developing" countries has become obsolete. It belongs to a national, capitalist, industrial subsystem of values that should be replaced with value systems that are more in line with an emergent ethics of global intelligence, based on second-order ethical principles (already present, as we have seen, in the age-old, planetary traditions of wisdom). From that standpoint, as we have noted in our conclusion to Chapter 6 above, there is no country that is more "developed" than another, so that all countries, geographical regions, and world cultures can bring their specific, invaluable contributions to human (self-) development.

An approach to knowledge and learning oriented toward global intelligence and planetary wisdom will require philosophical and scientific presuppositions entirely different from those of a power-oriented mindset. It does not presuppose that knowledge is power, but only that power produces certain forms of knowledge, which may become irrelevant or transfigured in other, non-power oriented, reference frames. It involves not only remapping traditional knowledge, as it is acquired, accumulated and transmitted by various academic disciplines, be they scientific or humanistic, but also generating new kinds of knowledge from a transdisciplinary and intercultural perspective.

By the same token, as we have noted in the conclusion of Chapter 4 of the present book, the quality and nature of the QR applications in a global learning and research framework will obviously depend on the quality and nature of the intercultural databases that they will draw upon and, above all, on the mentality and principles that will inform the collection and processing of such databases. It is here that the humanities (presently grievously neglected and/or relegated to a marginal position in our academic institutions by "bottom-line," utilitarian administrators) can bring their most decisive contribution. Intercultural and crossdisciplinary groups of researchers such as philosophers, cultural historians, anthropologists, environmentalists, sociologists, psychologists, educators, historians of science, linguists, literary scholars and many others, could compile intercultural data from a comparative perspective, delving into the systems of values and beliefs of various cultures, their philosophical, scientific, religious, and literary traditions, their specific economic, sociocultural, environmental, and legal practices, institutional arrangements, etc.

Such intercultural data, placed in a comparative perspective, but generated from the local viewpoint of each culture or subculture, whether large or small, and not from the so-called "objective" and "universally valid" perspective of mainstream Western science, would go a long way toward creating the local-global learning conditions that would lead to the adoption of the values and practices of global intelligence and planetary wisdom, based on a second-order ethics—again, the only kind of ethics that could truly work within a global reference frame. These are the values and practices that should infuse any DFO/FOR-based learning and research technological platform, programmed for the benefit of all (not just some) of our world communities.

Our most urgent task, then, would be to launch local-global, intercultural learning initiatives throughout the planet, such as we have envisaged in Chapter 8 above. What we need to learn or relearn in the first place is how to relate to each other and to our environment, including our machines, in mutually beneficial and enriching ways. There is a most urgent need to educate our world leaders and world populations in the spirit of global intelligence and planetary wisdom. This task can certainly be greatly facilitated by developing and using new ICTs such as the QR-based technological platforms, in the same spirit. It could also be greatly facilitated by a thorough exploration and understanding of the systems of values and beliefs that have produced the current socioeconomic, political and cultural institutions in various parts of the world, the ways in which these systems and subsystems have been interacting with each other throughout human history, and the best ways in which we can negotiate and resolve apparent and real conflicts among them, or consolidate and amplify their mutually beneficial feedback loops.

Above all, education itself and the very purpose and organization of our learning institutions must now be rethought and restructured within a global reference frame. A global perspective will lead to remapping the old disciplinary divisions and will generally call for new ways of educating the elites of tomorrow. Indeed, it will ultimately require that learning become a life-long process and extend well beyond formal education and certain age groups to all members of our local-global communities. Under the impact of life-long learning, these communities will ideally become genuine laboratories of cooperative, inter-cultural discovery and creativity.

Finally, an emergent ethics of global intelligence and planetary wisdom can best be grounded in a mentality of peace, defined not in opposition to war, but as an alternative mode of being, thinking, and acting in the world. This mentality has its own body of values and beliefs, emerging through intercultural research, dialogue and cooperation, and generates its own reference frames, organized on principles other than domination, cutthroat competition, and violence. It is such a mentality that can best nurture further human (self-) development and that should inform not only the ethical stance, but also all other aspects of future human activities, including the creation of new information and communication technologies.

For all of the foregoing reasons, we fervently hope that humanity will soon embark on this new evolutionary path and that future historians will look back to and characterize the 21st century as the Age of Global Learning.

Brief History of the Quantum Relations Principle

The idea of the Quantum Relation Principle (QRP) first arose in the late 1980s as Hardy F. Schloer's thought experiment of building a general, interdisciplinary model of machine-assisted analysis, encompassing such fields as psychology, sociology, physics and ideas of Eastern and Western philosophy. Schloer presented his idea in a 30-page essay on "The Quantum Relation Theory of Collective Human and Machine Thinking." The essay was initially inspired by the sudden stock market decline in October of 1987 and explored possible ways of understanding and foreseeing irrational market crashes before they occur.

The essay shed light on how human perception and thinking processes could be transported to a digital computer environment, using the relativistic approach of multiple observers within multiple frames of reference. It suggested that the Quantum Relations Principle could potentially be applied to the collective human-machine environments in order to archive the combined observing and thinking in a symbiotic way. Schloer's objective was to move local machine thinking to global, collective and cooperative grids of supercomputers that could observe and process in parallel as many observed facts as possible and understand them holistically and dynamically.

As the QRP was developed during the days of CompuServe Net, its ideas and application possibilities became greatly enhanced with the arrival of the commercial Internet in the early 1990s. Schloer started working on a commercially viable QRT solution and created several important innovations, charting the early road into cloud computing (see European Patent EP1126674B1 filed in December 1999)

In Spring of 2000, Dr. Philip A. Gagner, a very experienced information technology expert and intellectual property attorney from Washington, DC decided to join Schloer in the QR research and helped to push its further development in a fundamental way. Owing to Gagner's solid background in Artificial Intelligence, which he studied at MIT under Marvin Minsky, the "father" of the Artificial Intelligence concept, the QR applications in computer code began to evolve rapidly towards computing models of the actual world. Schloer provided the first functional framework through a first implementation of Cloud Computing, built between 1998 and 2001 in partnership with PatentPool Group in Munich, and over the ensuing decade Gagner began to fill it with functional details and operators, in order to bring QRP's multidimensional, parallel-space thinking to a still largely flat and linear world of computation at the dawn of the 21st century.

In Spring 2001, Dr. Mihai I. Spariosu, Distinguished Research Professor at the University of Georgia, Athens, USA became aware of the QRP at the Forum21 Think Tank conference in Paris, where Schloer gave a speech on machines that could think better than humans. Spariosu decided to join the research on QRP and helped to further define and develop the philosophical and practical implications of Schloer's ideas. He introduced the QRP and its many possible applications to a worldwide academic audience through his now often cited book, *Remapping Knowledge* (2006).

For the past decade, Schloer, Gagner and Spariosu have been working on strategies to make QRT a mainstream standard of intelligent global computing. Today QR-technological applications are being used in global consulting for both the public and the private sectors, as well as in advanced academic research. Through innovative theoretical thinking and EI-based machine implementations, Schloer, Gagner and Spariosu bring abstract, philosophical models of human and machine symbiotic learning and analyzing into the concrete world of problem solving and service applications.

Bibliographical References

Bak, Per. 1996. *How Nature Works*. Berlin and New York: Springer-Verlag.

Balas, E.A., Boren, S.A. (2000), "Managing Clinical Knowledge for Healthcare Improvement," in Yearbook of Medical Informatics, National Library of Medicine, Bethesda, MD: 65–70.

Bohm, David (1990), "A New Theory of the Relationship of Mind and Matter, in *Philosophical Psychology*. 3 (2): 271–286.

Capra, Fritjof. 1997. *The Web of Life. A New Synthesis of Mind and Matter*. London: HarperCollins.

Dyson, George B. 1997. *Darwin Among the Machines. The Evolution of Global Intelligence*. New York: Addison-Wesley Publishing Company.

Erickson, L.C., Wise, P.H., Cook, E.F., Beiser, A., Newburger, J.W. (2000), "The Impact of Managed Care Insurance on Use of Lower-Mortality Hospitals by Children Undergoing Cardiac Surgery in California." *Pediatrics*. 105(6):1271–8.

Gleick, James. 1987. *Chaos. Making a New Science*. New York: Penguin Books.

Gould, Stephen Jay and Niles Eldredge. 1977. "Punctuated Equilibrium: The Tempo and Mode of Evolution Reconsidered," in *Paleobiology* 3: 115–151.

Guttmann R. (1998), "Agent-Mediated Electronic Commerce: Issues, Challenges and Some Viewpoints," 2nd International Conference on Autonomous Agents, Minneapolis.

Hall, David L. and Ames, Roger T. 1987. *Thinking through Confucius*. Albany, N.Y.: State University of New York Press.

Heng, Michael Siam-Heng and Lim Tai Wei (2009), "Destructive Creativity of Wall Street and the East Asian Response" (Singapore: World Scientific), unpublished manuscript.

IOM (2001), "Crossing the Quality Chasm: A New Health System for the 21st Century" at https://iom.nationalacademies.org/Reports/2001/Crossing-the-Quality-Chasm-A-New-Health-System-for-the-21st-Century.aspx.

Jaspers, Karl (1953*), The Origin and Goal of History*. Translated by Michael Bullock. New Haven, CT: Yale University Press.

Mohd, Daud Bakar and Engku Rabiah Adawiah Engku Ali (2008), *Essential Readings in Islamic Finance* (Kuala Lumpur, Malaysia: CERT).

Nathin, Paul I. 2011. *Dr. Euler's Fabulous Formula*, Princeton, NJ: Princeton University Press.

Kauffman, Stuart. 1993. *The Origins of Order. Self-Organization and Selection in Evolution.* New York: Oxford University Press.

Lao Tzu. 1963. *Tao Te Ching.* Translated by D. C. Lau. London: Penguin Books.

Laszlo, Ervin. 1969. *System, Structure, Experience.* New York: Gordon and Breach.

–. 1973. *Introduction to Systems Philosophy.* New York: Harper.

Lovelock, James. 1988. *The Ages of Gaia. A Biography of Our Living Earth.* New York: W.W. Norton.

Macy, Joanna. 1991. *Mutual Causality in Buddhism and General Systems Theory: The Dharma of Natural Systems.* Albany, N.Y.: State University of New York Press.

Mandelbrot, Benoit. [1975] 1983. *The Fractal Geometry of Nature.* New York: Freeman.

Margulis, Lynn and Dorion Sagan. 1986. *Microcosm.* New York: Summit.

Margulis, Lynn. 1989. "Gaia: The Living Earth." Dialogue with Fritjof Capra. *The Elmwood Newsletter*, Berkeley, Calif., vol. 5, no. 2.

– and Dorion Sagan. 1995. *What Is Life?* New York: Simon and Schuster.

–. 1992. *Diversity of Life.* Berkeley Heights, N. J.: Enslow Publishers.

Maruyama, Magoroh. 1963. "The Second Cybernetics." *American Scientist* 51: 164–179.

–. 1974. "Paradigmatology and its Application to Cross-Disciplinary, Cross-Professional and Cross-Cultural Communication," Cybernetica 17: 135–156, 237–281.

Maturana, Humberto and Francisco Varela. [1970] 1980. *Autopoiesis and Cognition.* Dordecht: Reidel.

Milsum, John H., editor. 1968. *Positive Feedback: A General Systems Approach to Positive/ Negative Feedback and Mutual Causality.* London: Pergamon Press.

Naess, Arne. 1973. "The Shallow and the Deep, Long-Range Ecology Movement. A Summary," *Inquiry*, vol.16: 95–100.

–. 1989. *Ecology, Community and Lifestyle.* Translator David Rothenberg. Cambridge: Cambridge University Press.

Porter, Michael E. (2001), "Strategy and the Internet," in Harvard Business Review, March: 62–78.

Prigogine, Ilya and Isabelle Stengers. 1984. *Order out of Chaos.* New York: Bantam.

Richardson, George P. 1992. *Feedback Thought in Social Science and Systems Theory.* Philadelphia: University of Pennsylvania Press.

Samyutta Nikaya. The Connected Discourses of the Buddha (2002). Translated by Bhikkhu Bodhi. Boston: Wisdom Publications.

Schardt S., Mayer J. 2010. "What is Evidence-Based Medicine?" University of North Carolina Health Sciences Library. http://www.hsl.unc.edu/lm/ebm/whatis.htm.

Spariosu, Mihai I. (2015), Modernism and Exile: Play, Liminality and the Exilic-Utopian Imagination.

–. 2006. *Remapping Knowledge.Intercultural Studies for a Global Age.* New York and Oxford: Berghahn.

–. (2005) Global Intelligence and Human Development: Toward an Ecology of Global Learning. Cambridge, Mass.: MIT Press.

–. 1997. *The Wreath of Wild Olive: Play, Liminality and the Study of Literature.* Albany, N.Y.: State University of New York Press.

Stapp, Henry P. 1993. *Mind, Matter and Quantum Mechanics.* Berlin, Heidelberg and New York: Springer-Verlag.

–. 2011. *Mindful Universe: Quantum Mechanics and the Participating Observer.* Berlin, Heidelberg and New York: Springer-Verlag.

Stock. Gregory. 1993. *Metaman.* London: Bantam.

Thom, Rene. [1972] 1989. *Structural Stability and Morphogenesis: An Outline of a General Theory of Models.* New York: Perseus Publishing.

Von Bertalanffy, Ludwig. 1968. *General Systems Theory.* New York: George Braziller.

Wells, H. G.. [1938] 1994. *World Brain: H. G. Wells on the Future of World Education.* London: Adamantine Press.

Wiener, Norbert. [1948] 1965. *Cybernetics. The Science of Control and Communication in the Animal and the Machine.* Second edition. Cambridge, Mass.: The MIT Press.

Wilde, Oscar. 1954. *The Works of Oscar Wilde.* Edited by G. F. Maine. New York: E. F. Dutton.

Further Volumes:

Volume 8: Christian Høgel
The Human and the Humane
Humanity as Argument from Cicero to Erasmus

2015. 130 pages, hardcover
ISBN 978-3-8471-0441-4

Volume 7: Oliver Kozlarek (ed.)
Multiple Experiences of Modernity
Toward a Humanist Critique of Modernity

2014. 228 pages, hardcover
ISBN 978-3-8471-0229-8

Volume 6: Jörn Rüsen (ed.)
Approaching Humankind
Towards an Intercultural Humanism

2013. 300 pages, hardcover
ISBN 978-3-8471-0058-4

Volume 5: Marius Turda (ed.)
Crafting Humans
From Genesis to Eugenics and Beyond

2013. 197 pages, hardcover
ISBN 978-3-8471-0059-1

Volume 4: Christoph Antweiler
Inclusive Humanism
Anthropological Basics for a Realistic Cosmopolitanism

2012. 262 pages, hardcover
ISBN 978-3-8471-0022-5

Volume 3: Mihai Spariosu / Jörn Rüsen (eds.)
Exploring Humanity
Intercultural Perspectives on Humanism

2012. 295 pages, hardcover
ISBN 978-3-8471-0016-4

Volume 2: Stefan Reichmuth / Jörn Rüsen / Aladdin Sarhan (eds.)
Humanism and Muslim Culture
Historical Heritage and Contemporary Challenges

2012. 188 pages, hardcover
ISBN 978-3-89971-937-6

Volume 1: Longxi Zhang (ed.)
The Concept of Humanity in an Age of Globalization
2012. 233 pages, hardcover
ISBN 978-3-89971-918-5

V&R Academic

Verlagsgruppe Vandenhoeck & Ruprecht | V&R unipress

www.v-r.de